Sericulture Refresher

(For All the Competitive Examinations)

SERICULTURE REFRESHER

(For All the Competitive Examinations)

Kishore S. M.

Saddahalli (V), Palicherlu (P), Shidlaghatta (T), Chikkaballapura (D)
KARNATAKA - 562 105

Bharath K. B.

Maggula (V), Imangala (P), Virajpet (T), Kodagu (D),
KARNATAKA - 571 218

Vidyashree S.

Hosuru (V), Jathavara (P), Chikkaballapura (T), Chikkaballapura (D)
KARNATAKA - 562 101

NARENDRA PUBLISHING HOUSE
New Delhi-110 085

First Edition 2025

ISBN : 978-93-56514-92-8

Published by:

NARENDRA PUBLISHING HOUSE

Publisher and Distributor

C-21, Varun Apartments, Sector-9
Rohini, Delhi- 110085 (INDIA)
Ph: +91-11-45025794, +919717874875
Email: info@nphindia.com, nphindia@gmail.com
Website: www.nphindia.com

Printed in India

Composed & Designed by Palak Graphic Designer

Contents

Preface

Sericulture, the art and science of silk production, has a rich history spanning millennia. This book endeavours to unravel the intricacies of sericulture, catering to both the novice enthusiast and the seasoned practitioner. Attempts have been made to integrate a broad spectrum of information encompassing all facets of sericulture, including Historical Development in Sericulture, Mulberry Cultivation, Morphology, Anatomy and Physiology of Silkworms, Silkworm Seed Production, Silkworm Rearing, Pests and Diseases of Silkworms, Silk Reeling Technology, Vanya Sericulture, and Solved Model Papers incorporating recent scientific advancements.

The book "Sericulture Refresher" delivers the core subject in a concise, easily assimilative form designed to compete for ICAR JRF, SRF, ARS, AO examinations, interviews, *viva-voce* and other competitive exams.

The book would help to revise the entire course of Sericulture within a short time and recollect the vast knowledge they gained by reading different books on the subject. Besides, it will be an extremely useful document for students, teachers and all those interested in the subject of Sericulture.

The purpose of writing this book is to present the competitive exam aspirants with an objective form of learning in the subject. The bits kind of question-answers are framed very critically so that they cover the entire topic in a sequence. We believe the book will meet the need of aspirants and serve the purpose meant for.

We accept the responsibility for errors that you might come across while reading. Worthy suggestions for improvement of the present book from readers will be greatly appreciated.

2025

Kishore S. M.

Bharath K. B.

Vidyashree S.

Dedicated to ____________

All Mothers

About the Authors

Mr. Kishore S. M. is pursuing his Doctoral degree in Agricultural Entomology at KSNUAHS, Shivamogga, through ICAR. An alumnus of UAS, Bangalore, he holds a Master's degree from PJTSAU, Hyderabad (ICAR) and has qualified ASRB NET in 2023. To date, he has written 8 research and 17 review papers in national and international journals and authored 07 book chapters and 40 popular articles. He has participated in national and international conferences, webinars, workshops, trainings, EDP programmes and serves as a reviewer for several high-rated journals.

Mr. Bharath K. B. is an alumnus of the University of Agricultural Sciences, Bangalore, holding graduate and postgraduate degrees in Sericulture. He was awarded gold medals for his academic excellence during the 57th UAS-B Convocation. He has authored numerous publications, including book chapters, popular articles, research and review papers in journals of scientific repute. He has participated in national and international seminars, conferences, webinars and workshops. Currently, he is employed with ITC Limited - Agri Business Division in Rajahmundry, Andhra Pradesh.

Ms. Vidyashree S. serves as a Sericulture Extension Officer, Department of Sericulture, Government of Karnataka. She did her B.Sc. (Sericulture) from College of Sericulture, Chintamani. Thereafter, she completed her M.Sc. (Agri.) in Sericulture from University of Agricultural Sciences, Bangalore and has qualified the National Eligibility Test (NET). She has participated in various national and international, seminars, webinars, conferences and training programmes to enhance her knowledge and skills in sericulture.

Detailed Syllabus

Latest ICAR Syllabus for Entomology, Plant Pathology, Nematology and Sericulture JRF/SRF, ASRB NET and ARS Mains Examinations

ICAR-JRF (Entomology, Plant Pathology, Nematology and Sericulture)

UNIT-I: Importance of agriculture in national economy; basic principles of crop production; cultivation of rice, wheat, chickpea, pigeon-pea, sugarcane, groundnut, tomato, Cole crops, mango, grapes, banana, oilseeds other than groundnut, soybean and mustard. Major soils of India, role of NPK and their deficiency symptoms. Mendelian genetics; elementary knowledge of photosynthesis; respiration, and transpiration; Major cropping systems (rice-wheat cropping, crop rotations, mixed cropping); soil degradation-soil salinity and acidity and management; some aspects of post-harvest technology; varietal improvement; importance of Heterosis in crop production; crop protection principles in field and storage. Major insect pests and diseases of agricultural crops like rice, cotton and pulses; oilseed crops like groundnut, soybean and mustard; vegetables like tomato and cole crops; fruit crops like mango and banana and their management principles. Transgenic crops. Important rural development programmes in India; organizational set up of agricultural research, education and extension in India; Elements of statistics.

UNIT-II: Classification of animal kingdom up to class; distinguishing characters up to orders in class Insecta; general organization of an insect external morphology with special reference to lepidopteran larvae, coleopteran adults; and honeybee; metamorphosis and moulting; different physiological systems; insect- plant relationship; insect pests of agricultural and horticultural crops, and their stored/processed products, insect vectors of plant diseases-identification, biology, nature of damage, and their management tactics; and pests of household, medical and veterinary importance and their control; useful and beneficial insects like honeybee, lac insect, silkworm and pollinators; Nematode taxonomy, biology of important plant parasitic nematodes and their control; entomopathogenic nematodes, basic principles of insect and nematode pest management-cultural, biological, insecticidal, quarantine, and regulatory aspects; insecticide classification and insecticide resistance management; and insect protective transgenic crops.

ICAR-SRF and ARS Mains (Entomology, Nematology, Plant Pathology and Sericulture)

Entomology

Unit 1: Systematics - History and development of Entomology, Evolution of insects, position of insects in the animal kingdom, characteristics of phylum Arthropoda, structural features of important arthropod groups such as Trilobita, Chelicerata and Mandibulata, structural features of important classes of phylum Arthropoda viz. Arachnida, Crustacea, Chilopoda, Diplopoda and Hexapoda. Classification of insects up to order level, habits, habitats and distinguishing features of different Order and important Families.

Unit 2: Morphology - Body wall, its structure, outgrowths, endoskeleton, Body regions, segmentation, sclerites and sutures, Insect Colours. Head and head appendages, types of mouth parts, antennae, their structure and types. Thorax structure, thoracic appendages and their modification. Wings, their modification and venation, Abdomen; structure, abdominal appendages both in Pterygota and Apterygota. External genitalia, general structure and modification in important insect orders.

Unit 3: Embryology, internal anatomy and physiology - Embryonic and post embryonic development, types of metamorphosis, physiology of ecdysis. General features and types of larvae and pupae. Structure, function, physiology and modifications of Digestive, Circulatory, Respiratory, Reproductive, Nervous, Excretory systems, Endocrine system and Sense Organs. Insect food and nutrition; minerals, carbohydrates, proteins and amino acids, lipids, vitamins and their role in growth and development, artificial diets.

Unit 4: Ecology - Concept of ecology, Environment and its components-biotic and abiotic factors and their effects on growth, development, diapause, population structure and dynamics, distribution and dispersal. Principles of biogeography and insect's biodiversity. Assessment of diversity indices. Biotic potential and environmental resistance. Climate change and adaptations. Ecosystems, agroecosystem analyses, ecological niche, their characteristics and functioning. Intra and inter-specific relationship; competition, predator-prey and host-parasitoid interactions. Food chain, food web and trophic relations. Life table studies, population models. Arthropod population monitoring, pest forecasting. Causes of pest outbreaks.

Unit 5: Biological control - Importance and scope of biological control, history of biological control. Biocontrol agents - parasitoids, predators, insect pathogens and weed killers. Important entomophagous insect Orders and Families. Ecological, biological, taxonomic, legal and economic aspects of biological control, phenomena of multiple parasitism, hyperparasitism, superparasitism and their applied importance. Principles and procedures of using exotic biocontrol agents. Utilization of natural biocontrol agents: conservation, habitat management and augmentation. Mass multiplication techniques and economics. Effective evaluation techniques, Biocontrol organizations in world and India. Successful cases of biological control of pests. Use of biotechnological tools in enhancing the potentials of Bio-Control Agents.

Unit 6: Chemical control and toxicology - History, scope and principles of chemical control. Insecticides, classification and mode of action - Conventional and IRAC. Formulations of insecticides. Penetration of insecticides. Physical, chemical and toxicological properties of different groups of insecticides. rodenticides, insect growth hormones. Insecticide induced resurgence. Combination insecticides. Pesticide hazards and environmental pollution. Safe use of pesticides, precautions, first aid treatments and antidotes. Insecticides Act 1968, Functions of CIB & RC, registration and quality control of insecticides. Evaluation of toxicity, methods of toxicity testing, determination of LD 50, LT 50, RL 50 etc. Pesticide residues in the environment and their dynamics of movements, methods of residue analysis. Good laboratory practices. Pharmacology of insect poisons. Metabolism of insecticides; detoxification enzymes and their role in metabolism. Selectivity of insecticides insecticide resistance; mechanism, genetics and management of insecticide resistance.

Unit 7: Host Plant Resistance (HPR) - Principles of HPR. Basis of resistance (antixenosis, antibiosis, tolerance). Host plant selection by phytophagous insects. Biophysical and biochemical bases of defense against phytophagous insects. Genetics of Resistance: vertical resistance, horizontal resistance, oligogenic resistance, polygenic resistance. Biotype development and break down of resistance. Tritrophic interactions, induced resistance.

Breeding for insect resistant crops and evaluation techniques. Biotechnological approaches and development of transgenic insect resistant plants, its advantages and limitations. Case histories. Insect resistance to transgenic plants and its management.

Unit 8: Novel approaches in pest control - Behavioural control: semiochemicals pheromones-types and uses, advantages and limitations. Hormonal control: types and function of insect hormones, insect hormone mimics, advantages and limitations. chemosterilants, antifeedants, attractants, repellents; their types, method of applications, advantages and limitations. Genetic control: concepts and methods, case histories, advantages and limitations. Genetic improvement and genetic engineering of bio control agents. Pest management in organic agriculture. Pest management in precision agriculture.

Unit 9: Integrated Pest Management (IPM) - History, concept and principles of IPM. Components of IPM: Host plant resistance, cultural, mechanical, physical, chemical, biological, genetic and behavioural control etc. System approach, Agro-ecosystem and cropping system vs. IPM. Concept of damage levels- Economic threshold levels (ETL), Economic injury levels (EIL) and their determination. IPM strategies for field and horticultural crops. IPM case histories. Constraints and Strategies of IPM implementation. Plant quarantine laws and regulations.

Unit 10: Pesticide application equipment - Types of appliances: sprayers, dusters, fog generators, smoke generators, soil injecting guns, seed treating drums, flame throwers, etc. Types of nozzles, international classification, their uses, spray patterns, particle size, measurement, drift and non-target effects of pesticides. Maintenance of appliances. Aerial application-principles, guidelines, factors affecting the effectiveness, systems, advantages and disadvantages.

Unit 11: Pests of field crops and their management - Distribution, host range, biology and bionomics, nature of damage and management of arthropod pests of cereals, millets, nutricereals, oilseeds, pulses, fibre crops, green manures, sugarcane and tobacco. Pests of importance: locusts, termites, hairy caterpillars, cutworms white grubs and invasive alien pests. Vertebrate and molluscan pests.

Unit 12: Pests of horticultural crops and their management - Distribution, host range, biology and bionomics, nature of damage and management of arthropod pests of vegetables, fruits, plantation crops, spices, condiments, medicinal and aromatic crops, ornamentals, underutilized and exotic fruits. Pest management under protected cultivation. Pests of mushrooms. Vertebrate and molluscan pests.

Unit 13: Pests of stored products and their management - Principles of grain storage. Storage structures, bulk storage and bag storage, their merits and demerits. Grain drying methods and aeration. Storage losses, sources of infestation, factors influencing losses. Insect pests in storage, biology, and nature of damage. Non-insect pests (rodents, birds, mites) and their nature of damage. Management methods: Physical, Mechanical, Chemical, Biological, Behavioural, Legal and special storage methods. Microflora in storage environment and their control. Regulated and quarantine pests. Integrated management of storage pests.

Unit 14: Insect and mite vectors of plant diseases - History of vector pathogen interactions, important vectors of plant diseases, ecology of vector pathogen interaction Common insect and mite vectors viz., aphids, leaf hoppers, plant hoppers, whiteflies, thrips, psylids, beetles, weevils, flies, and mites and their relationship with the plant pathogenic fungi, bacteria, viruses, mycoplasma. Mechanism of pathogen transmission: Active mechanical transmission, biological transmission. Toxicogenic insects, mites and phytotoxemia. Some important arthropod vector transmitted diseases and their epidemiology in India. Management of vector and its effect on control of diseases. Role of climate change in vector borne diseases.

Unit 15: Honey bees and bee-keeping - History of bee-keeping. Honey bees and their economic importance. Bee products. Bee species, their behaviour, habit and habitats. Bee Keeping: bee pasturage, hives and equipment's, seasonal management. Bee poisoning. Bee enemies including diseases and their management. Quality analysis of honey. Pollinators and their role in production of various crops. Conservation of pollinators.

Unit 16: Silkworms and sericulture - Silkworm species, salient features, systematic position. Production techniques of mulberry, muga, eri and tasar silkworms. Nutritional requirements of silkworms. Sericulture: rearing house and appliances, silkworm breeds, principles of voltinism and moultinism, seed production and its economics. Different molecular approaches in developing silkworm breeds. transgenic silkworm Mulberry pests, diseases and their management. By products of sericulture and its value addition, uses in pharmaceutical industry. Enemies and diseases of silkworms and their management. Sericulture organization in India.

Unit 17: Lac insect - Biology, habit and habitats, lac products, uses. Host trees: pruning, inoculation, lac cropping techniques, and harvesting. Enemies of lac insect and their control

Unit 18: Helpful and useful insects - Pollinators and their role in production of various crops. Conservation of pollinators. Pollinators, insects as food, soil fertility improving agents, scavengers. Use of insects and insect products in medicines. Insects as bio-indicators. Usefulness of insects in scientific investigations

Unit 19: Statistics and computer application - Frequency distribution, mean, mode and median. Standard, normal, binomial and Poisson's distribution, Sampling methods and standard errors. Correlation and regression: Partial and multiple, tests of significance; *t, F, chi-square, Duncan's multiple range tests*. Design of experiments: Principles of Randomized block design, completely randomized block design, Latin square design, Split-plot designs. Probit analysis. Use of software packages like SPSS, SAS, etc. for the above tests and designs of experiments for analysis.

Plant Pathology

Unit 1: History and principles of plant pathology - Milestones in phytopathology with particular reference to India. Major epidemics and their social impacts. Historical developments of chemicals, legislative, cultural and biological protection measures including classification of plant diseases. Physiologic specialization, Koch's postulates. Growth, reproduction, survival and dispersal of plant pathogens. Factors influencing infection, colonization and development of symptoms.

Unit 2: Laboratory and analytical techniques - Preparation and sterilization of common media. Methods of isolation of pathogens and their identification. Preservation of microorganisms in pure culture. Methods of inoculation. Measurement of plant disease. Detection and Diagnosis of pathogens in seeds and other planting materials: Nucleic acid probes, Southern, Northern and Western hybridization, ELISA, ISEM and PCR. Nucleic acid probes,

Southern, Northern and Western hybridization, ELISA, ISEM and PCR. Laboratory equipment and their use: autoclave, hot air oven, laminar flow, spectrophotometer, electrophoresis, light and electron microscopy, incubator, ultracentrifuge, ELISA Reader, Freeze dryer, Nano drop, GC-MS, HPLC, Thermocycler.

Unit 3: Physiological and molecular plant pathology - Altered metabolism of plants under biotic and abiotic stresses. Molecular mechanisms of pathogenesis: elicitors, recognition phenomenon, penetration, invasion, primary disease determinant. Enzymes and toxins in relation to plant disease. Mechanisms of resistance, Structural and Biochemical defense mechanisms. R-Genes, Phytoanticipins. Phytoalexins. PR proteins, Hydroxyproline rich glycoproteins (HRGP). Antiviral proteins. SAR and ISR. HR and active oxygen radicals. Elementary genetic engineering. Management of pathogens through satellite, antisense - RNA. Ribozymes, coat protein, RNA interference, plantibodies, hypovirulence, cross-protection. Useful genes and promoters, plant transformation techniques, biosafety and bioethics.

Unit 4: Mycology - Classification of fungi (According to the Classification - Kirk *et al.*, 2008). Life cycles of important phytopathogenic fungi. Economic mycology, edible fungi and entomogenous fungi. Mycorrhizal associations. Cell organelles, their morphology, functions and chemical composition.

Unit 5: Plant bacteriology - Identification and classification of bacteria. morphology, ultrastructure and chemical composition of prokaryotic cell in relation to function. Growth curve, nutrition and auxotrophic mutants. Resting cells in elementary bacterial genetics and variability: transformation, conjugation, transduction. Biology of extra chromosomal elements: plasmid borne genes and their expression. Bacteriophages: lytic and lysogenic cycles. Prokaryotic inhibitors and their mode of action. Economic uses of prokaryotes. Morphology, biochemical characteristics, reproduction and life cycle of phytoplasma and other fastidious prokaryotes.

Unit 6: Plant virology - Nature, composition and architecture of viruses and viroids. Properties of viruses. Nomenclature and classification of viruses. Variability in viruses. Satellite viruses and satellite RNA. Mycoviruses and baculoviruses. Assay of plant viruses including biological, physical, chemical, serological and molecular methods. Conventional and biotechnological techniques used in detection and diagnosis. Behaviour of viruses in plants including infection, replication and movement. Histopathological changes induced by viruses in plants, inclusion bodies. Transmission of viruses: virus - vector relationships.

Unit 7: Plant disease epidemiology - Concepts in epidemiology. Development of disease in plant population. Monocyclic and polycyclic pathogens. Role of environment and meteorological factors in the development of plant disease epidemics. Survey, surveillance (including through remote sensing), and prediction and forecasting of diseases. Epidemic analysis and prediction models. Crop loss assessment: critical and multiple point models. Decision support system, cloud computing, GPS, GIS and GS in plant disease epidemiology.

Unit 8: Phanerogamic parasites and non-parasitic diseases caused by phanerogamic parasites and their management. Diseases due to unfavourable soil environment, drought and flooding stress etc. Nutritional deficiencies. Primary /secondary air pollutants and acid rain.

Unit 9: Fungal diseases of crop plants - Fungal diseases of cereals, millets, oilseeds, pulses, fruits, vegetables, plantation, fibre, spices, medicinal and ornamental crops with special reference to etiology, disease cycle, perpetuation, epidemiology and management. Postharvest diseases in transit and storage; aflatoxins and other mycotoxins and their integrated management.

Unit 10: Bacterial and viral diseases of crop plants - Crop diseases of cereals, pulses, oilseeds, sugar crops, vegetables, fruits, plantation and fibre crops caused by bacteria, viruses, viroids, phytoplasmas and other fastidious prokaryotes. Mode of transmission and pathogen vector relationships. Epidemiology and management.

Unit 11: Management of plant diseases - General principles of plant quarantine. Exotic pathogens and pathogens introduced into India. Sanitary and phytosanitary issues under WTO, TRIPS and PRA. Genetic basis of disease resistance and pathogenicity: gene for gene hypothesis; breeding for disease resistance. Production of disease-free seeds and planting materials. Seed certification. Chemical nature and classification of fungicides and antibiotics: their bioassay and compatibility with other agricultural chemicals; resistance to fungicides/ antibiotics; effect on environment. Spraying and dusting equipment's, their care and maintenances. Important cultural practices and their role in disease management, solarization, integrated disease management. Microorganisms antagonistic to plant pathogens in soil, rhizosphere and phyllo sphere and their use in the control of plant diseases; soil fungi stasis. Plant growth promoting Rhizobacteria. Biotechnology for crop disease management.

Plant Nematology

Unit 1: History and economic importance - History and development of nematology at national and international level. Position of nematodes in animal kingdom. Economic importance of nematology; distribution of economically important nematodes (*Pratylenchus*, *Radopholus*, *Hirschmanniella*, *Meloidogyne*, *Heterodera*, *Globodera*, *Rotylenchulus*, *Tylenchulus*, *Ditylenchus*, *Anguina*, *Aphelenchoides*, *Tylenchorhynchus*, *Helicotylenchus*, *Hoplolaimus*, *Scutellonema*, *Paratylenchus*, etc.). Beneficial nematodes. (entomopathogenic nematodes -*Steinernema* and *Heterorhabditis*), Parasites of insects (*Mermis*, *Agamermis*, *Romanomermis*). Entomopathogenic nematodes. Importance of saprophytes in organic matter recycling. Nematodes as indicators of pollution and toxicity. Predacious Nematodes.

Unit 2: Nematode taxonomy and morphology - Principles and concepts of taxonomy. Rules of nomenclature. Nematode phylogeny and systematics. Classification of soil and plant-parasitic nematodes and their relationships with other related phyla. Detailed classification of plant - parasitic nematodes up to generic level with emphasis on genera of economic importance. General morphology and anatomy of nematodes (cuticle, hypodermis, sense organs, pseudocoelem). Various systems: digestive, excretory, nervous, reproductive etc., developmental biology of nematodes.

Unit 3: Nematological techniques - Different methods of sampling for nematodes. Methods of extraction of nematodes from soil and plant material. Extraction of cysts nematodes. Microscopy - principles and types including electron microscopes. Methods of killing, fixing, preserving, staining, mounting and measuring of nematodes. Techniques for histopathology, Microtome techniques and culturing of nematodes - plant parasitic and entomophilic and saprophytic including axenic methods. Experimental techniques for proving pathogenicity, estimation of crop losses, nematicide screening, screening and evaluation for nematode resistance in crops. Molecular technique for nematode diagnostics. Techniques for mass culturing of entomopathogenic nematode antagonistic bioagents.

Unit 4: Nematode ecology - Ecological classification and distribution of nematodes. Mode of nematode dispersal. Adaptations to parasite mode of life. Soil as environment for nematodes. Effect of biotic and abiotic factors on nematode survival, activity and reproduction. Nematode population dynamics. Nematode-induced plant damage and modelling. Community analysis.

Unit 5: Plant-nematode relationships - Types of parasitism in nematodes. Classification of nematodes based on feeding habits. Nature of damage caused by various groups of plant parasitic nematodes and mechanisms involved. Pathotypes and races of root-knot, cyst and burrowing nematodes. Histopathological changes due to root-knot and cyst nematode infestation including syncytia, giant cell formation and their modification etc. Mechanism of nematode resistance and tolerance in plants and its assessment. Physiological, biochemical and molecular changes in plants due to nematode infections.

Unit 6: Nematode physiology and cytology - Chemical composition of nematodes. Principles of nematode physiology. Physiological functions of cell; organelles. Physiology of respiration, digestion, excretion, reproduction, growth and development. Physiology of moulting, hatching and nematode survival. Chemoreception in nematodes. Biology of root knot, cyst, Reniform, citrus, burrowing and lesion nematodes. Nematode as biological models - *Caenorhabditis elegans*. Cytological changes in plants due to infection including syncytia, giant cell formation and their modification etc.

Unit 7: Nematode pests of crops - Major nematode parasites and management in cereals (rice, wheat and barley), millets (sorghum, and maize), pulses (redgram, blackgram, greengram cowpea and chickpea),oilseeds (castor, groundnut and gingely), fibre crops (cotton and jute), vegetables (tomato, brinjal, bhendi, chilli and potato), cole crops (cabbage, carrot, cauliflower), beet root, sugarbeet, tapioca, tobacco, tuber crops (yam, dioscorea), fruits (banana, citrus, grapevine, guava, papaya, pomegranate), spices and plantation crops (turmeric, pepper, betelvine, arecanut and coconut), flower crops (crossandra, jasmine, carnation, rose, gerbera, chrysanthemum, eustoma and tuberose) and medicinal and aromatic plants, (medicinal coleus, geranium and patchouli), mushroom.

Unit 8: Nematode management - Principles and methods of nematode management - physical, cultural biological, chemical and legislative (plant quarantine), Nematicides (including those of biological origin) - history, classification, formulations, application and mode of action. Host resistance for nematode management. Resistance sources for different nematodes. Integrated nematode management. Nematode management in protected cultivation.

Unit 9: Interactions of nematodes with soil organisms- Importance of interactions (interrelationships) of nematodes with soil organisms. Interactions of nematodes with bacteria, fungi, viruses, mycorrhizae and other nematodes. Nematodes as vectors of viruses and other microorganisms.

Unit 10: Statistics - Frequency distribution. Measures of central tendency and dispersion: mean, median, mode, standard deviation etc. Population distributions: normal, binomial and Poisson. Correlations: regression, partial and multiple. regression, partial and multiple. Tests of significance: *t, F and Chi-square* and randomized block, Latin square and split-plot designs, their analysis and interpretation.

Sericulture

Unit 1: Mulberry crop production - Mulberry genetic variability and distribution, varieties. Soil suitability. Climatic requirements and constraints. Propagation methods. Planting methods, intercropping, population geometry, growth analysis. Soil fertility, fertilizer recommendation, integrated nutrient management, organic farming. Water management. Weed management. Problematic soils and management. Rainfed mulberry. Chawki garden maintenance. Training and pruning. Leaf quality, concept, assessment, preservation. Sericulture in integrated farming system. Resource management in mulberry crop production. Mechanization, machineries in sericulture. Pests, diseases and nematodes of mulberry and their management.

Unit 2: Mulberry and silkworm breeding – Mulberry - Germplasm, collection, conservation, evaluation and utilization. Choice of parental genotypes for selection and hybridization. Inter-varietal and interspecific hybridization and poly-cross. Clones, clonal selection. Breeding for quality. Genetic relationship between mulberry and silkworm genotypes, environmental interactions. Biotic and abiotic stresses. Selection of suitable host genotypes for stability and responsiveness across the environments. Selection for different situations like drought and salinity. Breeding for tree types. Selection of genotypes responsive to pruning. Innovative breeding approaches. Tissue culture/micropropagation of mulberry plants. Contribution of national and international institutes. Distinguishing characters of released varieties. Production of nucleus and breeders planting materials. Silkworm - Present status of silkworm breeding. Establishing a silkworm breeding programme. Introduction of exotic gene source. Inbreeding techniques, crossbreeding techniques for hybridization, different methods of mutation, convergent crossing, diallelic selection. Selection of good inbreed combiners. Selection for single trait and multi traits. Breeding for tolerance to pathogens, early maturity, high temperature. Post-cocoon characters – breeding for biochemical parameters and sex-limited characters. Recent advances in silkworm breeding.

Unit 3: Silkworm biology - Position of sericigenous insects in class Insecta. Silkworm integument, moulting process, exoskeleton. Structure and function of body regions. Head and its appendages, mouth parts, types of antennae, cephalic glands, mouth parts in silkworms and other insects. Thorax and its appendages, modification of legs in silkworms and other insects, type of wings and venation in *Bombycidae* and *Saturnidae*. Abdomen and its appendages, male and female genitalia in silk moths, male and female reproductive system, spermatogenesis, oogenesis, unusual types of development. Morphology and anatomy of eggs of silkworm. Diapause, physiology of diapause, biochemical changes, stages of development in diapausing and non-diapausing eggs, principles underlying breaking diapause and cold storage. Structure and function of silk gland, protein requirement in silkworm, composition of amino acids in silk gland, transamination, chemistry of fibroin, sericin and P25 (fibrohexamerin). Biosynthesis of silk, role of food supplementation on silk yield. Structure and function of digestive system, circulatory system, respiratory system, excretory system, endocrine system. Role of JH and anti-JH analogues/ phytohormones in cocoon production.

Unit 4: Silkworm protection - Insect pathology, history, concepts. Silkworm diseases, pathogenicity, kinds of infection, symptoms and pathologies associated with various diseases, classification of non-infectious and infectious diseases. Poisoning due to gases, tobacco poisoning. Nutritional and genetic diseases. Silkworm viruses, resistance of silkworm breeds against viral infection, role of antiviral and viral inhibitory factors, non-inclusion viruses. Miscellaneous flacherie diseases. Bacterial diseases, etiology, epizootiology, bacterial toxicosis, symptoms, management of bacterial diseases. Protozoan diseases, biodiversity, symptoms, detection, monitoring and management. Silkworm mycosis, types of fungal diseases, symptoms and management. Disinfection and hygiene, disinfectants, mode of action. Pests of silkworm, uzi fly - external morphology, biology, symptoms and management practices. Pests of grainage. Non-insect pests – mites, nematodes, rodents, symptoms and management practices.

Unit 5: Silkworm nutrition - Insect nutrition, nutritional requirements of silkworm larva, nutritional composition of mulberry leaves, feeding physiology of silkworm. Carbohydrate requirement, carbohydrate metabolism, lipid metabolism, synthesis of fatty acid in *Bombyx mori*. Amino acid requirement. Transamination. Silk proteins, physio-chemical properties, fibroin and sericin synthesis. Supplementation of nutrients for high silk yield. Role of water-soluble and lipid soluble vitamins. Growth factor. Role and requirement of minerals. Larval excretion. Dry matter economy. Intestinal flora of silkworm. Artificial diet, nutrient management through food supplementation.

Unit 6: Non-mulberry sericulture - Status of vanya silk industry in India. History and types of non-mulberry silkworms. Distribution and eco races of non-mulberry silkworms. Maintenance of germplasm of non-mulberry silkworms. Eri silkworm - morphology and voltinism, their host plants, castor for dual purpose. Tasar silkworm - food plants and cultivation practices, large scale plantations - *Terminalia arjuna* and *Terminalia tomentosa*, pests and diseases on food plants and their management. Muga silkworm - food plants and cultivation practices, pests and diseases on food plants and their management. Rearing of different non-mulberry silkworms. Pests and diseases of non-mulberry silkworms and their management practices. Rearing technology, economics of rearing, marketing system, silk reeling techniques. Recent advances in non-mulberry silk industry. Biotechnological approaches.

Unit 7: Silk reeling technology - Cocoon marketing, price fixation, sorting, mixing, drying and stifling. Cocoon cooking principle and methods. Types of reeling machines. Comparative output, advantages and disadvantages of different reeling devices. Reeling water quality. Silk throwing and weaving. Processes involved in winding, doubling, twisting, and rewinding for warp and weft hank making. Wet processing, degumming and bleaching by alkali and enzyme methods, dyeing different dyes, warping and weaving, printing

- block, hand screen tables. Grading of silk, standards, silk exchange. By-products and their uses.

Unit 8: Silkworm seed cocoon and egg production - Status and strategies of silkworm egg production. Morphology. Embryology. Biochemical changes in Eggs. Seed cocoon preservation, selection, sorting, sex separation, cocoon melting. Moth emergence, sexing, pairing and depairing, egg laying. Handling of eggs, embryological tools, diapausing and non-diapausing eggs. Acid treatment. Egg preservation schedule. Incubation/Black boxing technique. Bivoltine seed production. Seed crop monitoring, agencies. Seed Organization. Seed Acts, licensing procedures. Pests of grainage, pebrine diagnosis and management.

1

HISTORY AND DEVELOPMENT OF SERICULTURE

Ramayana Epic	:	The trousseau of Sita consisted of fine silk vestments of diverse colours, woollen clothing, furs, precious stones, princely ornaments and grand carriages of different kinds. The veil, bodice and body clothes are repeatedly mentioned in the epics Ramayana and Mahabharata. The garments used by the bride and groom at the wedding were made of silk fabric with swan motifs woven over them.
Mahabharata Epic	:	Silk fabrics were among the presents brought to King Yudhisthira by feudatory princes from the Himalayan regions after their conquest of the world.
Buddhist Jatakas	:	Silk fabric ('Koseyya' in Pali language) is mentioned as the famous rich fabric of Banaras.
2800 BC	:	Chin-Nong, one of the successors of Emperor Fo-Hi initiated the cultivation of mulberry in China
2640 BC	:	Queen Si Ling-chi (Xi Ling-chi or Lei-tzu) discovered the silk-making process and is renowned as the "Lady of the Silkworm" and "Goddess of Silk".
2200 BC	:	Authentic records of silk are found in the Chinese book, "Chronicles of Chou-king".
1600 BC	:	Silkworm Breeding started during the Shang period in China.

1200 BC	:	Chinese immigrants in Korea started sericulture, making Korea the first country to learn the secret of sericulture after China.
1100-770 BC	:	Poems and odes on silk are found in the book called "Song of the Western Zhone Period".
1027 BC	:	Well-organized silkworm breeding programme during the Chow dynasty.
1000-1077 BC	:	A necklace with a silkworm motif was unearthed in Baoji near Xian in the Shamshi province of China.
4th - 2nd Century BC	:	Trade of Chinese silk during the Mauryan period.
384-322 BC	:	Silkworm was mentioned in the book, "Historia Animalum" by Aristotle as "*Certain great worm bornlayn or spinner, which differ very much from other insects*".
321-296 BC	:	Kautilya's "Arthashastra" mentions 'Patrona', a variety of silk which was prominent in Magadha, Pundriya and Subarakhundya.
3rd Century BC	:	Introduction of Sericulture to Japan. Semiramis, General of Japanese Emperor Singre Kongo, invaded and conquered Korea and brought home invaluable wealth with captives, which included sericulturists. They began to practice sericulture in Japan. Emperor Ninken ordered his subjects to plant mulberry trees across the country.
206 BC - 8 AD	:	Silk weaving and printing attained a high degree of perfection during the Han dynasty and was considered the Golden Age of Sericulture in Chinese history.
150 BC	:	Silk shawls and woven silk fabrics found its popularity in the courts of Kings in Kashmir. The region was also known for its beautiful mulberry trees.

140 BC :	According to Western historians, sericulture spread to India from China through Kotan (Tibet).
138-118 BC :	Han Emperor Wu Ti established contact with the West through the "Silk Road" from China to Rome.
126 BC :	Silk from China and India was traded in Europe.
58 BC :	During the regime of Kanishka, silk was exported from India to Rome.
14 AD :	Measures were taken to restrict the import of silk to save the Roman Economy.
2nd Century AD :	Records of India shipping raw silk and silk fabric to Persia.
220—265 AD :	During the Wei dynasty in China, the government accepted tax payments in silk.
270 AD :	Roman Emperor Aurelian passed an order forbidding women to wear silk as it was worth gold and could cut down the import expenditure.
3rd & 5th Century AD :	A large number of Chinese technicians moved to Japan and initiated the silk weaving industry.
400 AD :	Sericulture spread to other parts of Asia, it began with a Chinese Princess marrying the King of Khotan (Tibet) who carried silkworm eggs and mulberry seeds concealing it in her headgear.
6th Century AD :	Roman Emperor Justinian I sent two Persian monks as Christian missionaries to Khotan to learn silk production techniques. The monks also smuggled silkworm eggs and mulberry seeds hiding them in their pilgrim staves.
540 AD :	Roman Emperor Justinian attempted to stop the drain of gold into China by deciding to smuggle the secret of silk culture out of China.

		The king offered a sufficient reward to Monks to smuggle silkworm eggs and mulberry seeds out of China, which they carried in a hollow stick, at the risk of their lives, for which death was the penalty.
606 - 647 AD	:	During the reign of the Vardhana dynasty King Harshavardhana, his court poet Banabhatta while describing the marriage of the King's sister, Rajyashri, mentions that the palace was arrayed in tentures flashing on every side like thousands of rainbows, dark silk, spider thread and shot silk, resembling slough of snakes, soft as unripe plantain fruit swaying a breath imperceptible except to the touch.
618 - 907 AD	:	During the Tang dynasty, China and Rome reigned at each end of the Silk Road.It was a productive era of sericulture in China, textile printing and colourful silkmaking were invented. Maximum traffic was observed on Silk Road and silk was smuggled to Constantinople (Istanbul).
650 AD	:	During the Taiko period, Japan's silk textile industry developed by adapting Chinese weaving and dyeing techniques. Japanese weavers also created intricate, multi-coloured Nishiki.People used to pay taxes in the form of silk goods.
740 AD	:	Sericulture was introduced into Spain by the Arab conquerors and it brought prosperity to Almeria, Murcia, Malaga, Saragosa and Valencia. Soon it began to export healthy silkworm seeds to Portugal and neighbouring countries. The silk industry declined in the second half of the century.
901 - 922 AD	:	The sericulture industry soon spread throughout Japan and gained importance.

11th Century AD	:	Sericulture was introduced to Southeast Asia towards the 11th century. Jawa was associated with silk production. Thai silk was hand-woven in Myanmar (Burma). Local tribes reared silkworms and dyed their clothes.
1147 - 1149 AD	:	Roger II, the Norman king of Sicily, invaded the Greek city of Thebes and forcibly brought skilled silk weavers back to Palermo, where they established the art of silk production in Sicily; this significantly contributed to the island's economic prosperity at the time.
1279 - 1368 AD	:	During the Yuan dynasty, satin became the most coveted silk fabric.
1340 AD	:	From Italy, the silk industry spread to France where silkworm rearing was introduced by French noblemen who brought in the silkworm eggs and the mulberry tree seeds.
1368 - 1644 AD	:	During the Ming dynasty, the Silk Road was virtually abandoned. Alexandria became the centre of silk distribution when oriental trade routes were changed to water routes. Planaria (Syria) became the clearing house for the silk brought from China and India, silk was also dyed here.
1500 AD	:	Japan's silkworm-rearing techniques reached the highest point producing a popular textile called Nishifin for the warriors called Samurais.
1500 - 1700 AD	:	The provinces of Guilan and Mazandaran on the Caspian Sea were important for sericulture during the Safavid dynasty of Iran.
1515 - 1541 AD	:	King Francis I of France is renowned as the "Father of silk industry in Europe", he notably brought Italian and Flemish weavers to Fontainebleau to produce exquisite tapestries.

1522 AD	:	Spanish conquistador Hernan Cortes introduced sericulture in Mexico.
1563 AD	:	Vittorio Amedeo II established a filature and later on, sericulture was spread to Milan, Como in Europe.
1589 AD	:	Henry IV opened the royal tapestry factory at Gobelins in Paris.
1603 - 1641 AD	:	During the reign of Ahom King Pratap Singh in northeast India, the golden yellow Muga silk had to be deposited to the royal treasurer for the king's household.
17th Century AD	:	Bengal was known for sericulture, and English agents reported large quantities of silk in 1621. The silk from Kasimbazar (near present-day Murshidabad) was so high quality that foreign traders, mainly Europeans, began calling it Bengal silk.
18th Century AD	:	Silk and copper became the most important commodities after rice in Japan.
Mid-18th Century AD	:	Italian reeling method was introduced in Indian factories to improve the quality of silk.
1801 AD	:	The jacquard loom was invented in France by Joseph-Marie Jacquard, a French weaver for weaving complex textiles, including silk
1771 AD	:	The Bengal Government under the East India Company obtained new silkworm breeds and mulberry varieties from China to improve the quality of cocoons.
1772 AD	:	The British established the first filature in Murshidabad of West Bengal.
1779 AD	:	The earliest reference to ericulture in India is documented in "Some Account of Silk in India, Especially of the Various Attempts to Encourage and Extend Sericulture" by J. Geoghegan.

1780 - 1790 AD	:	Tipu Sultan erstwhile ruler of Mysore and popularly renowned as "Father of the Karnataka Silk Industry" sent artisans to Bengal to learn the art of rearing silkworms and silk trade.
1786 AD	:	Tipu Sultan received letters from the Persian Gulf in the consul at Muscat to obtain the silkworms and experts.
1792 AD	:	The Bengal silk industry suffered a serious setback during the Revolutionary Wars.
1798 AD	:	The East India Company abandoned its silk projects and closed the silk farm established in Madras.
1813 AD	:	The decline of silk exports from India was observed because of Hoogly River changed its course and left it without its main means of transportation.
1826 AD	:	Funds were set apart to carry out experiments in mulberry cultivation, silkworm rearing and improvement of reeling in West Bengal.
1827 AD	:	An experimental silk filature was set up in Howrah.
1830 - 1850 AD	:	China, the ancient source, had bounced back into the centre of the market, and is the first place as a supplier particularly to France.
1840 AD	:	East India Company offered prizes under the British Commission for best-reeled silk in the array of Mysore and Kollegal.
1860 AD	:	An Italian company established a filature near Bangalore, imported silkworm races from Japan and tried to introduce crossbreed (Mysore race × Japanese race). However, failed due to muscardine disease.

1869 - 1890 AD	:	The epidemic of pebrine disease in Europe and the simultaneous opening of the Suez Canal were a boon to Japan which exported silk to the West and earned foreign exchange.
1869 AD	:	Ranbir Singh, King of Kashmir set up 127 silkworm-rearing houses.
1869 AD	:	Mysore, Bangalore and Kolar districts have accounted for almost the entire production of Mysore state.
1870 AD	:	Louis Pasteur rescued the silk industry by eradicating the deadly disease, which was later identified as Pebrine, by developing a preventive technique through simple microscopic examination of adult moths.
1870 AD	:	Louis Pasteur published his pioneering work on pebrine in "Etudes sur la Maladie des Vers a Soie" (Studies on Silkworm Disease).
1870 AD	:	The first International Sericultural Congress was held in Goritez, Italy.
1870 AD	:	The muscardine problem was widespread and resulted in continuous failure of production.
1878 AD	:	Indian sericulture industry received a setback due to the pebrine disease outbreak.
1880 AD	:	The first experiment with silkworm breeding and at the same time pebrine disease destroyed the whole sericulture industry in Europe.
1896 AD	:	JRD Tata established the Tata Silk Farm with a filature attached to it in Japanese pattern in Bangalore, with the help of Sri. K. Sheshadri Ayyar, the Diwan of Mysore. He employed Japanese people as managers and staff. While the Mysore government used their farm and Japanese staff to give training/teaching to local people.

Early 20th Century	:	Pebrine caused heavy losses to silkworm crops in France.
1900 - 1930 AD	:	East India Company set up six filatures. Trade centres at Surat in the west and Morulpatham in the east coast were opened.
1911 AD	:	The Economic Conference of Mysore had an Agricultural Committee and an Industrial and Commercial Committee, and it was decided that the former should deal with rearing and eradicating silkworm disease.
1912 AD	:	Sir. M. Visweshwaraiah, the appointed Diwan of Mysore State, spearheaded efforts to enhance sericulture. As part of his initiatives, he incorporated sericulture lessons into the curriculum of several primary schools. Additionally, eight trained schoolmasters from the silk farm were assigned to teach the subject.
1914 AD	:	The inception of the Department of Sericulture in Mysore state.
1914 AD	:	Government Silk Farm with the objective of research and development was established at Channapatna under Signor Washington Mari, an Italian expert in producing disease-free silkworm eggs.
1915 AD	:	H. Maxwell-Lefroy took up a position travelling across India and made a thorough survey as an 'Imperial Silk Specialist'. He along with E.C. Ansorge, magistrate and collector in Bihar and Orissa; and an amateur entomologist, wrote a "Report on an Inquiry into the Silk Industry in India" (1917) in three volumes.
1917 AD	:	Muscardine was again rampant in parts of the silk tract, contributing to a decline in the mulberry acreage in India.

1918 AD	:	The advisory council on silk appointed by the Imperial Institute London recommended the setting up of a sericultural research institute in India.
1919 AD	:	Services of Japanese expert Mr. Yonemura were obtained to study the combining abilities of the Mysore silkworm race with selected Chinese and Japanese Bivoltine races.
1920 AD	:	Hybrid silkworm eggs were widely introduced and utilised in Japan.
1920 - 1921 AD	:	Introduction of cross-breed silkworms in Mysore region.
1921 AD	:	The Government Silk Filature was established, subsequently, it became a joint stock company under the name Mysore Silk Filatures.
1925 AD	:	Mysore Domestic Basin, a new reeling machine was developed by N. Rama Rao, then Superintendent of Sericulture, Mysore State.
1925 AD	:	The first-ever 'Handbook of Sericulture' was published by Yonemura and N. Rama Rao.
1927 AD	:	The Mysore Silk Association was formed.
1929 AD	:	Automatic reeling machine of Katakura K-8 type was invented.
1932 AD	:	The Government Silk Weaving Factory was established in Mysore.
1934 AD	:	The Tariff Commission granted protection against cheap foreign silk imports.
1936 AD	:	The Government of Mysore started Spun Silk Mill as a joint stock company in Channapatna.
1940 AD	:	All India Sericulture Conference was held in Lucknow and the resolution was passed to establish a Central Sericulture Research Institute.

1940 AD	:	The hill amelioration station was established at BR Hills, Chamarajanagar District, Karnataka to maintain and multiply imported bivoltine and multivoltine silkworm races.
1942 AD	:	The Silk Conference was organised in Delhi.
1942 AD	:	To boost silk production, the Filature Expansion Scheme was implemented.
1942 AD	:	Silk Conditioning and Testing House was established in Mysore State.
1943 AD	:	Mysore Silkworm Disease Control Act was enacted.
1943 AD	:	Central Sericultural Research & Training Institute (CSRTI), Berhampore was established to provide research and development support for the promotion of the silk industry in the Eastern and Northeastern regions.
1947 AD	:	Dr. Syama Prasad Mukherjee, the then Minister of Industry and Supply introduced legislation in the Constituent Assembly of India to establish Central Silk Board.
1948 AD	:	The Seventh International Sericultural Congress was held in Ales, France. It was decided there to organise international meetings on sericulture regularly and it created the "Permanent Commission of International Sericultural Congresses".
1948 AD	:	"La Revue du Ver a Soie - Journal of Silkworms" journal publication started under the auspices of the Permanent Çommission of International Sericultural Congresses held in France.
1949 AD	:	Central Silk Board was constituted by the Government of India in April with Dr. Syama Prasad Mukherjee as its first chairman and N. Rama Rao as its vice-chairman under the Central Silk Board Act LXI of 1948 to empower the coordinated development of the silk industry.

1951 - 1956 AD	:	First Five-Year Plan - sericulture was included in "other village industries". Output of raw silk rose from 8.94 Lakh kgs at the beginning of the plan to 14.86 Lakh kgs at the end, accounting for a 66% increase.
1954 AD	:	Mysore Silkworm Seed (Control and Distributions) Act, 1952 came into effect.
1955 AD	:	The first International Technical Sericultural Conference met in Alès, France, adopted the proposal of transforming the "Permanent Commission of International Sericultural Congresses" into an inter-governmental organization, the "International Sericultural Commission (ISC)".
1955 AD	:	The Sericulture Research Institute was established in Channapatna with the assistance of the Central Government.
1956 AD	:	The Mysore Silkworm Diseases Control Act was enacted. Andhra Pradesh Silkworm Seed Act was passed.
1956 - 1961 AD	:	Second Five-Year Plan - Sericulture received an allocation of 379.25 Lakh, emphasising silkworm seed organisation and improvement of silk reeling.- Central Silkworm Seed Station was set up at Pampore in 1958. - All India Sericultural Training Institute was set up at Mysore.- Large-scale introduction of Improved Cottage Basins to improve the quality and output of raw silk.- Raw silk production increased from 14.86 lakh kgs to 15.13 Lakh kgs in 1960-61.
1957 AD	:	Silk Marketing Co-operative Society was established in Bengaluru.

1958 AD	:	Central Silkworm Seed Station, now named Central Sericultural Research & Training Institute (CSRTI), Pampore was established to entrust research and development of mulberry sericulture in North India.
1959 AD	:	Karnataka Silkworm Seed, Cocoon and Silk Yarn (Regulation of Production, Supply, Distribution and Sale) Act was enacted. It made provisions for the establishment of regulated markets at notified places.
1959 AD	:	India became a member country of the International Sericultural Commission (ISC) on 19th January.
1960 AD	:	China replaced Japan as the world's largest silk producer. Sericulture in Japan declined due to rapid industrial development, high labour controls and waning demand for traditional kimonos.
1960 AD	:	The International Sericultural Commission (ISC) was institutionalized on 8th August, after the ratification of the Instituting Convention, initially by four Member Countries namely, France, India, Romania and Yugoslavia with Mr. André Schenk (France) as its first Secretary General.
1961 -1966 AD	:	Third Five-Year Plan - Sericulture was assigned a separate place within the overall budget head of "Village and Small Industries". An outlay of 7.02 Cr was approved for developing the sericulture industry.- The Central Sericultural Research Institute was established at Channapatna in 1961.- Central Tasar Research Station was established at Ranchi in 1964, now Central Tasar Research and Training Institute (CTRTI).- Improved reeling and spinning equipment were introduced for Tasar and Muga cocoons.-

		On the export front, quality control inspection and standards were introduced in 1965.- "Central Silk Trade Mark" scheme was introduced in 1965.- Raw silk production increased from 15.13 Lakh kgs to 20.65 Lakh kgs, of which Karnataka's share was 12.05 Lakh kgs.
1961 AD	:	Government of Karnataka established 4 cocoon markets at Ramanagara, Chikkaballapura. Chamarajanagara and T. Narasipura.
1962 AD	:	"Indian Silk", a magazine on silk development with comprehensive coverage of market intelligence was launched by the Central Silk Board.
1963 AD	:	Central Sericultural Research Institute was relocated from Channapatna to Mysore and subsequently merged with the All-India Sericultural Training Institute to form the Central Sericultural Research & Training Institute (CSRTI), Mysore.
1964 AD	:	Basic Tasar Seed Station (now Central Tasar Silkworm Seed Station) was established at Lakha, Chhattisgarh.
1964 AD	:	Central Silk Board was importing raw silk on behalf of the state trading cooperation.
1968-69 AD	:	The interspecific hybrid of Tasar silkworm was developed by Manjeet S. Jolly at Central Tasar Research Station, Ranchi, the hybrid was *Antheraea proylei* (*Antheraea roylei* × *Antheraea perni*).
1969 - 1974 AD	:	Fourth Five-Year Plan - Plan envisaged to attain self-sufficiency to meet the country's demand for raw silk through increased productivity.- Raw silk production increased to 30.37 Lakh kgs.

1970 AD	:	Import of raw silk through the Central Silk Board was revoked.
1970's	:	Rearing of cross-breed silkworms started commercially in Karnataka.
1972 AD	:	Central Muga Eri Research Station was established at Titabar, Assam to provide R&D support to muga and eri silk industries in North East India (Assam, Manipur and West Bengal).
1972 AD	:	Regional Tasar Research Station on temperate oak tasar was established at Bhimtal, Uttarakhand.
1972-73 AD	:	Central Silk Board established the Raw Material Bank at Chaibasa, Jharkhand to bring stability to Tasar cocoon market conditions and prevent exploitation of tribals.
1974 - 1978 AD	:	Fifth Five-Year Plan - Programme for the introduction of bivoltine hybrids in traditional multivoltine states was initiated. - Central Silk Board established 11 grainages to hasten the bivotine introduction programme.- Price Stabilisation Scheme was introduced during 1977-78 to ensure fair economic returns to primary producers and the availability of raw silk to weavers and manufacturing exporters at a fair and steady price.- Raw silk production reached 37.11 Lakh kgs.
1974 AD	:	The Research Coordination Committee of Central Silk Board accepted the bivoltine hybrid of KA × NB4D2 as an approved bivoltine hybrid of Central Silk Board and was named 'Nandi' as the meeting took place in Nandi Hills.
1975 AD	:	National Silkworm Seed Organisation (NSSO) was established to cater for the need for quality mulberry silkworm seed in the country.

1975 AD	:	Manjeet S. Jolly published "Tasar Culture".
1975-76 AD	:	A separate bivoltine silkworm seed area was organised at Anekal, Karnataka.
1979 AD	:	Karnataka Government introduced 'Silk Exchange' to regulate the marketing of silk yarn.
1979 AD	:	"La Revue du Ver a Soie - Journal of Silkworms" journal started to be issued under a new name, "Sericologia".
1979 AD	:	Department of Sericulture was institutionalised by the Tamil Nadu Government with headquarters in Salem.
1980 AD	:	An International Centre for Training and Research in Tropical Sericulture (ICTRETS) came into existence at the premises of CSRTI, Mysore under the Indo-Swiss Technical and Scientific Co-operation Programme to provide advanced training in tropical sericulture.
1980-88 AD	:	World Bank assisted Karnataka Sericulture Project (KSP) was implemented in Karnataka state during the Sixth Five Year Plan.
1981 AD	:	National Silkworm Seed Project came into existence.
1981 AD	:	Karnataka State Sericulture Research and Development Institute (KSSRDI) was established in Talaghattapura, Bengaluru under the World Bank assisted Karnataka Sericulture Project as an applied research and development institute, with the main object of addressing research and development aspects of the Department of Sericulture, GOK, with special emphasis on the field problems.
1982 AD	:	India stood as the largest net exporter of mulberry silk with net exports of 530 Tonnes.

1982-83 AD	:	University of Agricultural Sciences, Bangalore started B.Sc. (Sericulture) degree programme under the 10+2+4 scheme.
1982 AD	:	Sher-e-Kashmir University of Agricultural Sciences and Technology of Kashmir (SKUAST-K) began offering sericulture courses.
1983 AD	:	Central Silk Technological Research Institute was set up in Bangalore to tackle issues related to post-cocoon technology.
1983 AD	:	Indian Silk Export Promotion Council (ISEPC) was established to promote and develop the export of natural silks, silk blends and finished products.
1983 AD	:	Muga Seed Development Project (MSDP) was launched in Guwahati.
1985 AD	:	The Seventh Five-Year Plan stated that employment generated through sericultural activity during the Sixth Five-Year Plan was 1.32 million persons.
1985-86 AD	:	An Inter-State Tasar Project (ISTP) was implemented in eight states to promote tasar production in organised *Terminalia arjuna* block plantations.
1986 AD	:	India rose to second position in silk production after the continued decline in Japan's silk production.
1987 AD	:	Central Muga Research & Training Institute was established at Lahdoigarh, Jorhat to provide research and development support to the muga silk industry.
1987 AD	:	The USA became the main market for Indian silk followed by West Germany and the UK, all together comprising 59% by value.

1987 AD	:	The Silkworm Seed Technology Laboratory (SSTL) was established in Kodathi, Bangalore by the Central Silk Board.
1987-88 AD	:	Indian raw silk exports reached 2200 Tonnes at a growth rate of 12.7% per annum.
1989 AD	:	India during its Seventh Five-Year Plan launched World Bank assisted National Sericulture Project (NSP 1989 to 1998) in 5 traditional and 12 pilot states aiming to increase mulberry raw silk production to 15000 Tonnes by 1994-95.
1991 AD	:	Japanese Government assisted JICA projects were launched in three phases from 1991 to 2007.- First Phase (1991 - 1997): Bivoltine sericulture technology development project.- Second Phase (1997 - 2002): Project of promotion and popularising the practical bivoltine sericulture technology.- Third Phase (2002 - 2007): Project for strengthening extension system for bivoltine sericulture.
1991 AD	:	Central Sericultural Germplasm Resources Centre (CSGRC) was established at Hosur for the conservation and utilization of seri-biodiversity in the country.
1993 AD	:	Seri-biotechnology Research Laboratory (SBRL) was established at Kodathi, Bangalore in collaboration with the World Bank under the National Sericulture Project.
1994 AD	:	"Year of Women in Sericulture" in India.
1994 AD	:	Central Silk Board and All India Silk Machinery Manufacturers Association organised the first-ever silk machinery exhibition at Bangalore "Silk Tech India - 1994".

1995 AD	:	Central Silk Board introduced the Race Authorisation System under the CSB Act of 1948 Section 8(1) to promote the development of silk industry by selecting high yielders for commercial exploitation.
1995 AD	:	College of Sericulture was started by the University of Agricultural Sciences, Bangalore at Chintamani, Chikkaballapura District.
1996 AD	:	Andhra Pradesh State Sericulture Research and Development Institute (APSSRDI) under the World Bank assisted the National Sericulture Project at Kirikera, Hindupur.
1999 AD	:	Central Muga Research & Training Institute was renamed Central Muga Eri Research & Training Institute (CMER&TI) with a mandate to serve as the apex research and development institute for Muga and Eri.
2004-05 AD	:	Catalytic Development Programme (CDP) scheme was initiated during the Ninth Five Year Plan to focus on the holistic development of the sericulture industry in different states of the country.
2005 AD	:	Muga Seed Development Project (MSDP) was renamed as the Muga Silkworm Seed Organization (MSSO).
2011-12 AD	:	Tamil Nadu Agricultural University started a four-year B.Sc. (Sericulture) degree programme.
2013 AD	:	The International Sericultural Commission (ISC) headquarters has been shifted from Lyon, France to Bangalore, India.
2014-15 AD	:	College of Sericulture was started under Assam Agricultural University at Titabar, Assam.

2017 AD	:	The Government of India launched a Central Sector Scheme "Silk Samagra" for three years 2017-2020 for the overall development of sericulture in the country.
2019 AD	:	The 6th edition of the Asia-Pacific Congress of Sericulture and Insect Biotechnology (APSERI - 2019) organised by the International Sericultural Commission (ISC) was held at Mysore, India.
2021 AD	:	Muga Silkworm Seed Organization (MSSO) was renamed as Muga Eri Silkworm Seed Organization (MESSO).
2021 AD	:	"Silk Samagra-2" was launched as an integrated scheme for the development of silk industry synergising the efforts of the State Governments and other implementing agencies to improve the quality, productivity and production of raw silk, besides generating employment opportunities, particularly in the rural areas during 2021-2026.
2022 AD	:	The recent 26th International Sericultural Commission Congress was held at Cluj Napoca, Romania.
2024 AD	:	The Platinum Jubilee celebration of the Central Silk Board was held in Mysore on 20th September.

The earliest archaeological evidence of silk in China comes from the Liangzhou Neolithic site of Qianshanyang dating to 2570 BC. Until recent times it was believed that the silk from a bead thread at Nevasa in Maharashtra dated to 1500 BC was the earliest-ever archaeological evidence of silk in India. The recent survey and observation of the excavated findings of Harappa and Chanhu-daro sites reveal the finds of silk with several other copper ornaments dates somewhere between 2450 and 2000 BC and their SEM imaging further divulges that the silk is of *Antheraea* sp. and *Philosamia* sp.

2

MULBERRY CULTIVATION

1. Mulberry is a **perennial** crop.
2. Place of origin of mulberry – **lower slope of the Himalayas (China)**.
3. The geographical position of mulberry growing countries practising sericulture lies **north of the equator** (only exception is Brazil).
4. The diploid somatic cell chromosome number of the genus *Morus* is **2n = 28**.
5. The highest ploidy of mulberry in nature is **2n = 22x = 308**.
6. The natural tetraploid species of mulberry is ***Morus laevigata*** (2n = 4x = 56).
7. The complete nucleotide sequence of the chloroplast genome of the mulberry (*Morus indica* cv. K2) is **158,484 base pairs (bp)**.
8. Mulberry inflorescence is called **catkin**.
9. The floral formula of male flower in mulberry (Family: Moraceae) is ⊕, ♂, $\mathbf{P_{2+2}}$, $\mathbf{A_4}$ (⊕ - actinomorphic, ♂ - male, P_{2+2} – two sets of two tepals in perianth, A_4 – androecium with four unfused stamens).
10. The floral formula of female flower in mulberry (Family: Moraceae) is ⊕, ♀, $\mathbf{P_{2+2}}$, $\mathbf{\underline{G}_{(2)}}$ ($\underline{G}_{(2)}$ - gynoecium with two fused carpels and superior ovary).
11. Mulberry fruit is **sorosis**.
12. The viability of fertile mulberry seeds is **3 months**.
13. Ideal planting material for mulberry plantations is **saplings**.
14. Saplings are **rooted cuttings**.
15. Age of saplings ideal for low or high bush plantations is **100-120 days**.
16. Age of saplings ideal for tree plantation is **240 days**.

17. Ideal soil for nursery establishment is **loamy soil**.
18. General nursery bed size, keeping working convenience in consideration is **10 × 4 ft** or **300 × 120 cm**.
19. Nursery bed of 300 × 120 cm accommodates **180 cuttings** (spacing 8 × 20 cm) required for low or high bush plantation.
20. Nursery bed of 300 × 120 cm accommodates **100 cuttings** (spacing 10 × 30 cm) required for tree plantation.
21. The age of shoots to prepare stem cuttings must be **6-8 months old**.
22. Ideal cutting size – **15 cm long, 10-15 mm diameter (pencil thickness) with 3-4 healthy buds**.
23. Cuttings preparation machine or shoot cutting machine was developed by **CSRTI, Mysuru**.
24. To enhance rooting, cuttings are treated with **NAA/ IBA** (200ppm aqueous solution of NAA/IBA with 90% ethyl alcohol – 6 hours).
25. To prevent fungal infection, cuttings are dipped in **0.2% Bavistin solution** (10-15 min.).
26. The most common disease in nursery is **leaf spot**.
27. To control fungal diseases in nursery, **0.1% Carbendazim (Bavistin)** spraying is recommended.
28. To control thrips in nursery, **0.1% Dimethoate (Rogor)** is recommended.
29. To control mites in nursery, **0.1% Oxydemeton-methyl (Metasystox)** is recommended.
30. The commercial establishment to raise mulberry saplings on a large scale and make them available to farmers is called **Kissan Nursery**.
31. The ideal soil pH for establishing mulberry plantation is **6.5 to 7.0**.
32. The ideal temperature, relative humidity and sun shine duration required for optimum growth and sprouting of buds in mulberry are **24-28 °C, 65-80 % and 5-12 hours per day**, respectively.
33. One-third of the mulberry area in southern and central India are under **rainfed** conditions.
34. The soil moisture stress under rainfed condition is experienced when rainfall is less than **700 mm**.

35. Under rainfed conditions, the high yielding mulberry variety recommended for red loamy soil is **S-13**.
36. Under rainfed conditions, the high yielding mulberry variety recommended for black cotton soil is **S-34**.
37. High-yielding mulberry variety recommended for temperate rainfed conditions – **Goshoerami**.
38. Ideal land for mulberry cultivation is **flat or slightly slopy**.
39. Main field plantation should be undertaken during **commencement of monsoon**.
40. Under rainfed conditions, the method of planting adopted is **pit system**.
41. For rainfed mulberry, the recommended dose of N:P:K fertilizer for the first year is **50:25:25 kg/ha** respectively.
42. For rainfed mulberry, the recommended dose of N:P:K fertilizer from second year onwards is **100:50:50 kg/ha** respectively.
43. Under rainfed conditions, the leaf yield is **10-12 MT/ha/year** (third year onwards).
44. Mulberry, without significant reduction in leaf yield can last for **12-15 years**.
45. The number of plants per unit area is **planting density**.
46. The number of plants in one hectare under 30 × 30 cm spacing is **1,11,111** plants.
47. The number of plants in one hectare under 30 × 60 cm spacing is **55,555** plants.
48. The number of plants in one hectare under 30 × 90 cm spacing is **37,037** plants.
49. The number of plants in one hectare under 60 × 60 cm spacing is **27,777** plants.
50. The number of plants in one hectare under 60 × 90 cm spacing is **18,518** plants.
51. The number of plants in one hectare under 90 × 90 cm spacing is **12,345** plants.

52. The number of plants in one hectare under 150 × 150 cm spacing is **4,444** plants.
53. Number of plants in one hectare under (60 × 90) × 150 cm spacing is **13,888** plants.
54. Mulberry in North India (temperate regions) is grown as **scattered tree** with the recommended spacing of **240 × 270 cm** (1,543 plants/ha).
55. Higher the plant density, **lower** is the leaf quality.
56. The process of systematically removing certain shoots in order to give mulberry plants proper shape and size, and also to stimulate bud sprouting is called **pruning**.
57. Pruning of shoots at the shoot base on the main stump is called **fist form pruning**.
58. Pruning of shoots leaving 2-3 buds in the shoot is called **non-fist form pruning**.
59. More number of shoots are produced in **non-fist form pruning**.
60. If the shoots are cut/harvested at the base (within 15-30 cm from ground), it is called **bottom pruning** or **Kolar system of pruning**.
61. If the shoots are cut at around mid-height (60-70 cm from ground), it is called **middle pruning**.
62. If the shoots are cut at the top confining to the green softwood portion, it is called **top pruning**.
63. To prevent widening of the crown and narrowing of space between the plants due to repeated prunings, **rejuvenate pruning** is recommended.
64. Pruning just below the first forking on the main stump is called **rejuvenate pruning**.
65. For improved quality of leaves, the number of shoots recommended is **8-10** per plant.
66. Irrigation water requirement for mulberry is **1.5 to 2 acre inch** per irrigation.
67. The ideal method of irrigation is **drip irrigation**.
68. Percentage water saving in drip irrigation compared to furrow irrigation is **30-40 %**.

69. Under irrigated conditions, the recommended dose of N:P:K fertilizer in shoot harvest system (5 harvests) is **300:120:120 kg/ha/year** (5 split doses).

70. Under irrigated conditions, the recommended dose of N:P:K fertilizer in shoot harvest system (5 harvests) for V-1 mulberry variety is **350:140:140 kg/ha/year** (5 split doses).

71. Under irrigated conditions, the recommended dose of N:P:K fertilizer in leaf harvest system (6 harvests) is **280:120:120 kg/ha/year** (6 split doses).

72. Under irrigated conditions, the recommended dose of N:P:K fertilizer for chawki garden is **225:150:150 kg/ha/year** (8 split doses).

73. Under irrigated conditions, the recommended dose of N:P:K fertilizer for hilly areas is **250:100:100 kg/ha/year** (5 split doses).

74. Farm yard manure (FYM) recommendation for mulberry garden is **20 MT/ha/year**.

75. Farm yard manure (FYM) recommendation for chawki mulberry garden is **40 MT/ha/year**.

76. Under irrigated conditions, the leaf yield is **35-60 MT/ha/year**.

77. In hilly terrain, the leaf yield is **25 MT/ha/year**.

78. Cheapest and best means mulberry leaf harvesting for silkworm rearing is **shoot harvesting**.

79. Mulberry leaf preference of young age or chawki silkworms – **tender, highly succulent and carbohydrate rich feed**.

80. Mulberry leaf preference of late age silkworms – **coarser leaf with less moisture and more fibre content**.

81. The quality of mulberry leaf is determined by **thickness of the cuticle, thickness of the leaf blade and number of idioblasts**.

82. The superior quality mulberry leaf has **thinner cuticle, thin leaf blade and less number of idioblasts**.

83. Most suitable mulberry variety for chawki rearing is **S-36**.

84. To maintain soil fertility and productivity in mulberry gardens for sustainable productivity, **integrated nutrient management (INM)** is followed.

85. An efficient way of waste management by reducing the C:N ratio, bulkiness and increasing the nutrient content is **composting**.
86. Earthworm species generally utilised in vermicomposting – ***Eudrillus eugineae, Eisenia foetida* and *Perionyx excavates***.
87. The criteria for choosing leguminous crops for green manuring are **short duration and bio-mass production capacity**.
88. The most suitable green manuring crop for red soil areas is **Sunhemp (*Crotalaria juncea*)**.
89. The most suitable green manuring crop for black soil areas is **Dhaincha (*Sesbania aculeata*)**.
90. Recommended seed rate for green manuring crops in mulberry gardens is **20-22 kg/ha**.
91. Azotobacter biofertilizer saves nitrogenous by **50%**.
92. Azotobacter biofertilizer is a bacterial preparation of live ***Azotobacter chroococcum* cells** with suitable carrier material.
93. 'Seri Azo', commercially available Azotobacter biofertilizer is developed by **CSRTI, Mysuru**.
94. Azotobacter biofertilizer recommendation for the mulberry garden is **20 kg/ha/year**.
95. 'Seriphos', phosphate solubilizing biofertilizer (*Bacillus megatherium* cells + carrier material) is developed by **KSSRDI, Bengaluru**.
96. 'Seriphos' biofertilizer recommendation for mulberry garden is **5 kg/ha/year**.
97. VAM inoculation saves phosphatic fertilizers by **75%**.
98. VAM inoculum recommendation for the mulberry garden is **1000 kg/ha**.
99. Beneficial VAM fungi for mulberry – ***Glomus mosseae* and *Glomus fasciculatum***.
100. Recommended VAM fungi spore density is **20 spores/ gram dry soil**.
101. IPNMS (Integrated Plant Nutrient Management System) model – **soil source → mineral fertilizers → organic sources → composting → vermicomposting → green leaf/ green manuring → biological sources → biofertilizers**.

102. A sustainable farming system called organic farming was advocated by **Mokichi Okada (1935)**.

103. The mulberry leaf lamina turning greenish-yellow to yellowish-white and to white leaving midrib and veins green (young → old leaves) is the characteristic symptom of **manganese (Mn)** deficiency.

104. The chlorotic conditions in mulberry with scorchy yellow to brown discolouration with darkened veins is the characteristic symptom of **zinc (Zn)** deficiency.

105. A multi-nutrient formulation for correcting nutrient deficiencies in mulberry developed by CSRTI, Mysuru is **Poshan**.

106. The leaf-based plant growth regulator that enhances photosynthate assimilation in plants and increases leaf yield is **triacontanol (TRIA)**.

107. Triacontanol (Vipul/ Triacontanol-cf) recommendation for the mulberry garden is **250 ml/ha/harvest** (2 split doses).

108. Triacontanol increases mulberry leaf yield by **15-20%**.

109. The foliar formulation known to increase mulberry foliage and improve the commercial characteristics of silkworm is **Seriboost**.

110. Seriboost increases mulberry leaf yield by **23%**.

111. Plant growth regulator formulation (PGR) developed by CSRTI, Berhampore for foliar application is '**Microzyme-B**'.

112. Anti-transpirant developed by CSRTI, Berhampore for water-stressed conditions is '**Jal Sanjivini**'.

113. Liquid formulation developed under the assistance of CSRTI, Mysuru for improvement of soil fertility, mulberry and silkworm productivity is '**Ankur**'.

114. The most obnoxious and prevalent dicot broad-leaved weed in mulberry is ***Parthenium hystrophorous***.

115. Major perennial monocot grassy weeds are ***Cynodon dactylon* and *Cyperus rotundus***.

116. A sustainable and eco-friendly method of perennial weed control is **digging**.

117. Post-emergence, non-selective, systemic and translocative herbicide recommended in mulberry gardens is **glyphosate** (@ 0.71% concentration, 600 litre spray solution per hectare).

118. The cropping pattern having no distinct rows between mulberry and intercrops is called **mixed cropping**.

119. The cropping pattern where mulberry and intercrops are planted in distinct rows is called **row intercropping**.

120. The cropping pattern where mulberry and intercrops are grown in separate strips is called **strip intercropping**.

121. The cropping pattern where a second intercrop is sown after the first intercrop in the interspace is called **relay intercropping**.

122. The cropping pattern where mulberry is grown with few other intercrops of varying heights or mulberry itself as an intercrop is called **multi-tier cropping**.

123. The shade tolerant variety of mulberry is **Sahana**.

124. The mulberry grown under shade possesses **less carbohydrates and proteins but more of inorganic nitrates**.

125. Maintenance of a hectare mulberry garden requires **800 man-days/year**.

126. Mechanization in mulberry gardens can reduce overall labour requirement to **1/10th** proposition.

127. The plantation method developed by CSTRI, Mysuru to completely mechanise mulberry cultivation and provide scope for intercultural operations is **3M plantation**.

128. An insect pest known to cause 'Tukra' symptoms (bunchy top or rosetting) in mulberry is **pink mealybug (*Maconellicoccus hirsutus*)**.

129. Mealybug incidence is higher in **summer**.

130. Predatory biocontrol agents against pink mealybug are ***Cryptolaemus montrouzieri*** **and** ***Scymnus coccivora***.

131. Exotic parasitoids effective in controlling polyphagous papaya mealybug (*Paracoccus marginatus*) are ***Acerophagus papaya, Pseudleptomastix mexicana*** **and** ***Anagyrus loecki***.

132. Mulberry leaves having white streaks later developing into silvery blotches is the characteristic symptom of **thrips (*Pseudodendrothrips mori*)** infestation.

133. Predatory biocontrol agents against thrips are **ladybird beetles (*Menochilus sexmaculatus* and *Scymnus coccivora*), *Chrysoperla* sp. and *Orius* sp.**

134. Thrips infestation is severe during **summer**.

135. The deposition of waxy secretion in a characteristic spiralling pattern is the symptom of **spiralling whitefly (*Aleurodicus dispersus*)**.

136. Parasitoids effective in controlling polyphagous spiralling whitefly are ***Encarsia guadeloupae* and *Encarsia haitiensis***.

137. Spiralling whitefly egg predator is ***Axinoscymnus puttarudriahi***.

138. Parasitic wasps against long-tailed mealybug, *Pseudococcus longispinus* are ***Anarhopus sydneyensis*, *Hungariella peregrina* and *Anagyrus fusciventris***.

139. 'Hopper burn' is a result of **jassid (*Empoasca flavescens*)** infestation.

140. Mulberry pest responsible for upward curling of leaves (cup-shaped) is **jassid (*Empoasca flavescens*)**.

141. Minor occasionally serious pest especially during winter season is **jassid (*Empoasca flavescens*)**.

142. Effective parasitoid against black scale insect (*Saissetia nigra*) is ***Metaphycus helvolus***.

143. The most prevalent mulberry sap sucker in hilly regions is **soft scale insect (*Megapulvinaria maxima*)**.

144. Water jetting technology is effective in the management of **sucking insects**.

145. Major defoliator pest known to cause extensive damage to mulberry is **leaf webber (*Diaphania pulverulentalis*)**.

146. Severe defoliator pest of mulberry in temperate region is ***Glyphodes pyloalis***.

147. Peak period of leaf webber infestation is **November to February**.

148. Target area of leaf webber infestation in mulberry plants is the **apical portion**.
149. Egg parasitoid against leaf webber – ***Trichogramma chilonis*** (1 Trichocard/acre/week)
150. Larval parasitoid against leaf webber – ***Bracon brevicornis*** (200 adult wasps/acre).
151. Pupal parasitoid against leaf webber – ***Tetrastichus howardii*** (1 pouch/acre).
152. One life cycle of *Diaphania pulverulentalis* is **17-24 days**.
153. Sporadically important pest prevalent in mulberry especially near vegetable growing regions is **cutworm (*Spodoptera litura*)**.
154. A pheromone trap used to attract and kill male cutworm moths – **Spodolure**.
155. Egg parasitoid against Bihar hairy caterpillar (*Spilosoma obliqua*) – ***Trichogramma chilonis*** (4 Trichocards/acre/week).
156. An occasional pest that spreads to other fresh branches and plants through wind by hanging on silken thread secreted by itself is **Noctuid moth caterpillar (*Tiracola* sp.)**.
157. Nocturnal occasional endemic serious pest of mulberry is **May-June beetle (*Holotrichia serrata*)**.
158. Girdling of the young or green stem and subsequent wilting are the characteristic symptoms of **stem girdler (*Sthenias grisator*)**.
159. Termite (*Odontotermes* sp.) attack on mulberry plants is prominent in **rainfed** conditions.
160. Molluscicide recommended for Great African snail (*Achatina fulica*) infesting mulberry – **Metaldehyde**.
161. Common management practise for black slug (*Laevicaulis alte*) is soil application of **lime powder** (20 kg/ac).
162. Snail and slug belongs to the phylum **Mollusca**.
163. The life stage of red spider mite (*Tetranychus ludeni*) causing damage to mulberry foliage is **both nymphs and adults**.
164. Mulberry foliar disease most prevalent in Southern India during monsoon and winter is **leaf spot**.

165. ‘Shot hole’ is a characteristic symptom of **leaf spot** disease.

166. Leaf spot disease in mulberry foliage is caused by ***Cercospora moricola***.

167. The appearance of white powdery patches on the lower surface and the corresponding upper surface develops chlorotic lesions due to the infection of ***Phyllactinia corylea* (fungus)**.

168. An obligate parasite and micro-cyclic black rust causing organism is ***Cerotelium fici***.

169. Rust fungus disperses through air by **uredospores**.

170. Mulberry leaf blight (fungus) is caused by ***Alternaria alternate/ Fusarium pallidoroseum***.

171. Mulberry leaf blight (bacteria) is caused by ***Pseudomonas syringae* pv. *mori* (= *Xanthomonas compestris* pv. *mori*)**.

172. Isolated irregular brown coloured patches are the preliminary symptom of **fungal leaf blight**.

173. Irregular water-soaked patches leading to curling and rotting of leaves is the symptom of **bacterial leaf blight**.

174. Favourable condition for bacterial leaf blight development – **higher temperature and humidity**.

175. The red/yellow colour triangular mark on the pesticide packaging denotes **higher** degree of toxicity.

176. The blue colour triangular mark on the pesticide packaging denotes **moderate** degree of toxicity.

177. The green colour triangular mark on the pesticide packaging denotes **negligible** degree of toxicity.

178. Pesticides solution required foliar spray in one-acre mulberry garden is **180 litres**.

179. The disease characterised by wilting of the saplings in the apex and progressing downwards, leading to death of saplings is **die-back (*Botryodiplodia theobromae*)**.

180. Dark discolouration of the bark and rooting od of cuttings near the collar region is due to ***Phoma* sp. (collar rot)**.

181. Secondary source of disease infection in case of nursery is **infected stem cuttings**.

182. *Trichoderma pseudokoningii* based bioformulation for nursery diseases, developed by CSRTI, Mysuru is '**Nursery-Guard**'.

183. Mulberry plants showing stunted growth, marginal chlorosis-necrosis of leaves and formation of galls on the roots are the characteristic symptoms of **root-knot nematode (*Meloidogyne incognita*)**.

184. 'Nemahari' is a plant-based formulation developed by CSRTI, Mysuru for the management of **root-knot nematode**.

185. Recommended intercrop in root-knot infested mulberry garden is **marigold (*Tagetus patula*)**.

186. 'Bionema' is a bioformulation developed by CSRTI, Mysuru to manage root-knot nematode has inoculum of ***Pochonia clamydosporia*** bioagent.

187. Mulberry root rot causing organisms – ***Fusarium solani, F. oxysporum, Botryodiplodia theobromae* and *Macrophomina phaseolina***.

188. Sudden withering of plants and defoliation of leaves starting from bottom of the branch and progressing upwards is the characteristic symptom of **root rot disease**.

189. The plant-based formulation developed by CSRTI, Mysuru for mulberry root rot disease management is '**Navinya**'.

190. Formulation involving Dithane M-45 integrated with bioagent *Trichoderma harzianum* developed by CSRTI, Mysuru is '**Raksha**'.

191. The board spectrum environment-friendly formulation developed by CSRTI, Mysuru for control of root rot disease in mulberry is '**Rot-fix**'.

192. Root disease complex is caused by the association of **root-knot nematode and root rot fungi**.

193. Bio-consortium developed by CSRTI, Mysuru against root disease complex is '**Bio-mix**'.

194. 'Bio-mix' involves bioagents ***Trichoderma harzianum* and *Trichoderma viride***.

195. The herbal formulation developed by CSRTI, Mysuru for the control of major foliar and soil-borne diseases is '**Chetak**'.

3

MORPHOLOGY, ANATOMY AND PHYSIOLOGY OF SILKWORMS

1. The pH value of haemolymph of the silkworm is **6.3 to 6.5.**
2. A change of constant amplitude in the membrane potential of a neuron that is triggered by a depolarization of the dendrite is called as **action potential.**
3. A displacement of the embryo within the egg is called **blastokinesis.**
4. A female moth of silkworm lays about **400 to 600 eggs.**
5. A female specific yolk protein that is synthesized by the fat body and taken up by the oocyte through receptor-mediated endocytosis is called as **vitellogenin.**
6. A glucagon-like hormone which brings about increase in blood trehalose by mobilization of fat body glycogen is **Hypertrahalocaemic hormone.**
7. A heterogeneous mixture of small peptides secreted by different neurosecretory cells of brain and cells of corpora cardiaca is called **Activation harmone.**
8. A large globular protein that comprises the thick myofilaments of muscle **Myosin.**
9. A pair of glands in the retro cerebral complex present close to the brain and receiving the two pairs of nerves from the NSCs of the brain is called **Corpora Cardiaca.**
10. A pair of small ovoid glandular bodies situated just beneath the brain and closely associated with corpora cardiaca is **Corpora Allata.**
11. A small peptide that causes contraction of the longitudinal muscles of the proctodeum of the hindgut is called **Proctolin.**
12. A space between the presynaptic neuron and the post synaptic neurons is called **Synaptic & Synapse.**

13. A special sac in the female reproductive system that receives, stores and releases sperms to fertilize the eggs it passes by in the oviduct **Spermatheca.**
14. A state of arrested development observed in insects exposed to unfavorable environmental conditions at which normal life is impossible for the organism is called **Diapause.**
15. A type of compound eye used by diurnal insects in which each ommatidia is surrounded by pigment cells is known as **Apposition eyes.**
16. A type of ovariole in silkworm, where nurse cells accompany the oocytes within the follicle and supply it with nutrients is **Polytropic ovarioles.**
17. A type of sperm mostly produced by male lepidopteran insects that lack nucleus. **Apyrene sperm.**
18. A unit of muscular contraction between the two Z-lines, containing the myofibrils is called as **Sarcomere.**
19. According to Muller (1828), corpora allata is also called as **Visceral Ganglia.**
20. Aggregation of numerous neurons and their nerve fibers is called as **Ganglion.**
21. Among the fatty acids, **Linoleic & Linolenic Acids** hold the highest nutritive value and are regarded as essential fatty acids for silkworm.
22. Among the haemocytes, which is the largest in size is **Oenocytoid.**
23. Among the silkworm races, the maximum number of silk gland cells are reported in **European races.**
24. An accessory cell that surrounds neurons and provides them nourishment and insulation is called **Glial cells.**
25. An enzyme hydrolysing dietary fats into fatty acid and glycerol is **Lipases.**
26. An ovariole in which there are no nurse cells is **Panoistic.**
27. Apical portion of the insect sperm which breaks down the egg membrane to allow the sperm to penetrate for fertilization is **Acrosome.**
28. At the anterior end of the egg, there is a small opening through which the spermatozoa enter the egg is called **Micropyle.**
29. At the point where the left and right anterior ends of the silk glands join, there is a pair of glands called **Lyonnet's and Filippi's Glands.**

30. Bombyxin in silkworm is produced by **Prothoracicotropic Hormone.**
31. Carbohydrate ingested from diet is mainly consumed in the body to supply energy, while the remains can be synthesized into storage carbohydrate such as **Glycogen.**
32. **Cellulose, Sugar, Isositol and Vit-C** acts as gorging stimulants of silkworm.
33. Chitin is soluble in **Sodium hypochlorite.**
34. **Cholestrol** can be turned into α - ecdysone which is the precursor of β -ecdysone.
35. Chordotonal organs consist of a single unit or group of similar units called **Scolopidia.**
36. **Citral, linalyl acetate, linalool** acts as feeding attractants for silkworm.
37. Cuticular sclerotization is also called as **Tanning.**
38. Each ommatidium contains a hard-crystalline cone from an intracellular secretion of the semper cells is **Eucone type.**
39. Each spiracle is set in a sclerotized cuticular plate called **Peritreme.**
40. Each trachea ends in a microscopic tube called **Tracheoles.**
41. Each trachea ends in a microscopic tube which penetrates the tissues is called **Tracheoles.**
42. Each vas deferentia becomes enlarge to form a sac known as **Seminal vesicle** in which sperms are stored.
43. Ecdysone was structurally identified by **Butenandt and Karlson.**
44. Fat body is a loose aggregation of **Mesodermal cells.**
45. Generally, most insect neurons are **Monopolar.**
46. **Goblet cells** excrete potassium, calcium and large inorganic molecules from haemolymph into the midgut lumen.
47. Hormone which is absent during pupal-adult metamorphosis is **JH.**
48. Hormones which are able to enter the cell and bind cytosolic and nuclear receptors are called **non-polar hormones.**

49. Hormones which cannot pass through the cell membrane, initiating the synthesis of second messenger molecules that carry the message inside the cell are called **Polar Hormones.**

50. How many hours after oviposition germ band formation is completed: **24hrs.**

51. How many types of haemocytes are reported in silkworm haemolymph is **5.**

52. In bivoltine and multivoltine races where the occurrence of diapause can be manipulated by environmental conditions is called **Facultative Diapause.**

53. In membrane depolarization which resulted due to the opening of the ionic channels of the post synaptic receptors due to binding of the transmitter molecules at the chemical synapse, the membrane becomes more permeable to **$Na^{\pm}$.**

54. In **non-hibernation** races colour of the eggs does not change till it reaches to blue egg stage (one day prior to hatching).

55. In silkworm, an enzyme which converts protein into amino acid is: **Proteases and Proteolytic enzymes.**

56. In silkworm, digestion of which carbohydrate is not possible is **Cellulose.**

57. In silkworm, the amount of uric acid excreted is about **85%.**

58. In silkworm, the haemolymph enters the dorsal vessel through **Ostia.**

59. In silkworm, which type of silk gland cells are responsible for silk yield are **Glandular cells.**

60. In some insects all energids do not migrate to the periphery to form the blastoderm but some of them lie behind within the yolk and are called **Vitellophages & yolk cells.**

61. In the silkworm haemolymph the main form of carbohydrate is **Trehalose** contributing 90% of total carbohydrate in haemolymph.

62. In which order of insect's tracheal system is lacking and integument serves as the function of gaseous exchange is **Collembola, Protura and Dipteran larvae.**

63. In which stage of silkworm haemolymph protein level is maximum **larvae.**

64. Insect brain is also called as **Supraoesophageal ganglion.**

65. Insect diet with all components completely known is **Meridic Diet.**

66. Insects in which embryonic development is completed within the female and gives birth to larvae or nymph instead of laying eggs is called as **Viviparity.**

67. Late age silkworms prefer **Mature leaves** which contain less water and more fibre provide the needed 'biting stimulus' and the balance of nutrients meet a totally different kind of growth and development.

68. Major component of silkworm integument is **Chitin.**

69. Malpighian tubule is derived from the major cell types of the principle cells from **Ectoderm and stellate cells from Mesoderm.**

70. **Morin** is a substance present in mulberry leaf to initiate feeding response in silkworm.

71. **Morin, beta sitosterol and Isoquercitrin** acts as biting-inducers of silkworm.

72. Moth urine is reddish brown, with some particular odors because of the presence of the **Riboflavin.**

73. Neurosecretory cells are indeed the source of the brain's endocrine effect was demonstrated by **Wigglesworth.**

74. Normally insect nerve axons have a resting potential of about is **-70mV.**

75. On the basis of functional spiracles, which type of respiratory system is found in silkworm is **Peripneustic.**

76. **Ovarioles** are functional units of ovary.

77. **Panoistic** are type of ovarioles in which nutritive cells are absent.

78. **Phagostimulants** are required to be present in the food not only to initiate a feeding response but also to sustain it till the end.

79. Pharynx, oesophagus, crop and proventriculus are the parts of **Foregut.**

80. Phenolic compound produced a hard-brown material, involved with X-linking protein molecules and hardening is **Sclerotins.**

81. **Plasmatocytes** plays a major role in phagocytosis and encapsulation.

82. Presence of excretory tubules at the junction of mid and hindgut were first reported by the Italian scientist **Marcello Malpighi** in 1669.
83. Protein component of the cuticle that is soluble in water is **Arthropodins.**
84. Proteinaceous secretions of the male accessory glands that enclose the sperms and protect it during transit to the female is **Spermatophore.**
85. Receptors which are sensitive to body movements are called as **Proprioceptors.**
86. Receptors which are sensitive to humidity are called as **Hygroreceptors.**
87. Sericin protein is secreted from which region of the silk gland is **Middle region.**
88. Sesquiterpenes in insects are produced by **JH.**
89. Silk gland excrete excessive amino acids and water from the silkworm which otherwise causes a disease known as **Aminoacidemia.**
90. Silk gland is developed during which stage of silkworm development: **Embryonic stage**.
91. Silk proteins is soluble in warm water is **Sericin.**
92. Silk-producing apparatus is a paired organ consisting of modified: **Labial glands.**
93. Silkworm cuticle is composed of **Chitin and Proteins.**
94. Silkworm grows naturally when fed on the diet without alanine, glycine, serine, cysteine or tyrosine so these kinds of amino acids are called **non-essential amino acids.**
95. Silkworm hacmolymph lacks **Haemoglobin.**
96. Silkworm nervous system is **Aminergic.**
97. Small intestine, colon and rectum are the parts of **Hindgut**.
98. Socket producing cell is called **Tormogen cells.**
99. Some fat body cells act as a storehouse for the excretory substances, particularly the uric acid granules called **Urate cells.**
100. Source of neuropeptide hormones in insects is **Corpora Cardiaca**.
101. **Spiracle** is the paired aperture in the integument that serves as the opening of the tracheal system to allow gaseous exchange.

102. Spiracles are also called as **Stigmata.**
103. Stomodeum and proctodaeum are **Ectodermal** in origin.
104. Structure of Malpighian tubules was discovered in 1669 by **Marcello Malpighi.**
105. Synthesis and deposition of yolk in the oocyte is called as **Vitellogenesis.**
106. Taenidia is absent in **Tracheoles and Air sacs**.
107. The alimentary canal of the silkworm is divided into **3regions.**
108. The amino acids needed by silkworm but they cannot synthesize or the amount of synthesis is very less and they must ingest from food are called **Essential amino acids.**
109. The amount of silk spun by spinning larva of the univoltine and bivoltine races is about **0.5gm/larvae.**
110. The aperture which opens externally on the body surface is called **Spiracle.**
111. The association between the sensory and motor neurons are called as **Inter neurons.**
112. The average number of silkworm eggs per gram is about **2000.**
113. The basic number of Malpighian tubules in silkworm is **6.**
114. The body cavity of insects is called the **Haemocoel** and the blood, the **Haemolymph.**
115. The body cavity through which insect blood bathes the tissues directly is called: **Haemocoel.**
116. The body form assumed during a particular stadium is called **Instar.**
117. The casting off old cuticle is called as **Ecdysis.**
118. The cellular building blocks of the insect nervous system are **Neurons.**
119. The chemical formula of chitin is **$C_8H_{13}O_5N$.**
120. The circulatory system in silkworm is **Open type.**
121. The cocoon spinning movement in silkworm is controlled by **Ventral nerve cord.**
122. The colour of the haemolymph in insects is yellow-green due to the presence of **Insectoverdine.**

123. The columnar and goblet cells are found in **Midgut.**
124. The contraction phase of the heartbeat results from the contractions of the muscles in the heart wall which start posteriorly and spread forward as a wave **Systole.**
125. The contribution of haemolymph towards the body weight of silkworm is **25%.**
126. The developing embryos receive nourishment through a special structure formed from follicle cells, that absorbs nutrients from "milk" secreted by the female accessory glands is called **Pseudo placental viviparity.**
127. The development of an embryo from an unfertilized egg nucleus is called **Gynogenesis.**
128. The development of an embryo into adult is called as **Metamorphosis.**
129. The digestive enzyme which converts disaccharide sucrose and maltose into sugar is known as **Invertase.**
130. The disc of protein within a muscle cell that separates the sarcomeres is called **Z-line.**
131. The distal ends of the Malpighian tubules bath freely in the haemolymph of haemocoel in **Basic nephridial system.**
132. The egg may either be fertilized or develop parthenogenetically is called **Obligatory parthenogenesis.**
133. The egg of silkworm has an outer protective covering is called **Chorion.**
134. The eggs of silkworm are **Centrolecithal.**
135. The embryo develops within the female but when the egg hatches, the larva is retained in the modified genital chamber and feeds from secretion of the female accessory gland is called **Adenotrophic viviparity.**
136. The embryonic stage which involves formation of the germinal layers is commonly referred as **Gastrulation.**
137. The endocrine function of corpora allata was described by **Wigglesworth 1934.**
138. The endocrine function of ecdysial glands was reported by **Fakuda 1940.**
139. The entire body of the embryo produces pigments turns black so that egg looks light blue through the chorion, after which embryo starts hatching next day and this is known as **Body pigmentation stage & Blue egg stage.**

140. The enzyme present in the synapse that breaks down acetylcholine to acetate and choline is **Acetylcholinesterase.**
141. The excretory system in silkworm is composed of two structures, and the **Malphigian tubules and Rectum.**
142. The expansion phase when haemolymph enters the heart, results from relaxation of the heart assisted by the elastic filaments that support the heart is known as **Diastole.**
143. The first artificial diet for silkworm was developed by **(Fakuda, 1960).**
144. The first evidence for the existence of hormone in insects was reported by **Bataillon.**
145. The functional unit of the ovary is called **Ovarioles.**
146. The glandular cells of corpus cardiacum secrete an octapeptide of molecular weight of about 850 daltons that is **Adipo kinetic hormone.**
147. The heart rests in the expanded condition and this phase are called **Diastasis.**
148. The hemocoel is divided into three compartments or sinuses **Dorsal pericardial sinus, Central perivisceral sinus and Ventral perineural sinus.**
149. The hind part of the brain responsible for labral nerves innervating the labrum is known as **Tritocerebrum.**
150. The hormone responsible for the hardening and darkening of cuticle is called **Bursicon.**
151. The hormone responsible for moulting **Ecdysone.**
152. The hormone secreted by median neurosecretory cells and involved in the temporal organization of eclosion behaviour in *Bombyx mori* and other Saturnid insects is **Eclosion Hormone.**
153. The hormone which acts on the Malpighian tubule to stimulate urine production in **Diuretic Hormone.**
154. The hormone that increases fluid reabsorption by acting on the hindgut is **Antidiuretic Hormone.**
155. The insect central nervous system consists of the **Brain and ventral nerve cord.**
156. The interval between ecdysis or intermolt period is called **Stadium.**

157. The intromittent organ of copulation in the male is **Aedeagus.**
158. The largest region of the silk gland is **Posterior Region.**
159. The larvae of silkworm are **Eruciform.**
160. The larvae take their food into the buccal cavity, drive the food backward by **Peristaltic action** of pharyngeal muscles.
161. The lepidopteron larvae are technically known as **Caterpillar.**
162. The major protein in egg yolk in silkworm is **Vitellin.**
163. The metabolic process that breaks down ingested organic molecules and releases energy, captured by ATP and waste products is called as **Catabolism.**
164. The metabolic production of protein, carbohydrates and fats from ingested food is called: **Anabolism.**
165. The moulting fluid is capable of digesting **Endocuticle.**
166. The movements of the embryo taking place from the posterior to anterior pole of the egg are **Anatrepsis.**
167. The movements taking place from the ventral to dorsal surface of the egg are **Katatrepsis.**
168. The muscle type found in silkworm is **Striated muscle.**
169. The newly hatched larvae of silkworm are called as **Kengo and silk ants.**
170. The normal steady membrane potential in an unstimulated neuron is called as **Resting Potential.**
171. The number of haemocytes in silkworm ranges from **1000 to 2000 cells per mm^3.**
172. The number of layers in silkworm procuticle is **2.**
173. The number of spiracles present in silkworm are **9 Pairs**.
174. The olfactory receptors are present on the antennae of male moths of silkworm that respond to the female sex pheromone is **Sensilla Trichoidea.**
175. The optically active structure of a light receptive sensillum consisting of the individual rhabdomeres of adjacent retinula cells is called **Rhabdom.**

176. The origin of the silk gland is **Ectodermal.**
177. The ovaries lie freely within the fat body and disperse eggs within the hemocoel and are surrounded by a group of follicle cells, the trophamnion, that feeds the oocyte from maternal tissues is called **Hemocoelous viviparity.**
178. The part of the brain associated with the compound eyes is known as **Protocerebrum.**
179. The part of the neuron that carries information to other cells is **Axon.**
180. The part of the neuron that is specialized for reception of information is **Dendrite.**
181. The peristaltic action is stopped during **Moulting.**
182. The peristaltic frequency is **26-35** moves per minute for 5th instar larvae.
183. The peritrophic membrane is found in **Midgut.**
184. The photoreceptors in silkworm are **compound eyes and dorsal ocelli.**
185. The pigment present in the serosal cells is attributed to one of the amino acids **Tryptophan.**
186. The posterior region of the silk gland is responsible for producing the silk: **fibrion.**
187. The precursor for ecdysteroid synthesis is **cholesterol.**
188. The primary storage organ of lipids in silkworm is **Fat body.**
189. The primordial male germ cell is called **Spermatogonia.**
190. The principal metabolic storage cells of the fat body are **Trophocytes.**
191. The process by which cuticle becomes stiffer and hard is called **Sclerotization.**
192. The production of several embryos from a single egg is called **Polyembryony.**
193. The pupae in which the appendages are firmly pressed against the body are called **Obtect.**
194. The pupae in which the appendages are free of any secondary attachment to the body are called **Exarate.**
195. The pylorus valve can regulate the movement of the digested food from **Midgut into Hindgut.**

196. The ringed appearance of the trachea to prevent collapsing is called **Taenidia.**
197. The sense of smell that is mediated by chemical stimuli in a gaseous state at low concentration is called as **Olfactory or smell chemoreceptors.**
198. The sense of taste that is mediated by chemical stimuli acting with liquids at high concentration is called as **Gustatory or contact chemoreceptors.**
199. The sense organ of insects which respond to the sense of touch and current of air is called as **Mechanoreceptors.**
200. The senses lacked by the silkworm larvae are **Sight & hearing.**
201. The sensory hair-producing cell is called as **Trichogen cell.**
202. The sensory organs of insects are **Mechanoreceptors, Chemoreceptors and Photoreceptors.**
203. The sensory stimulation by chemicals is called as **Chemoreceptors.**
204. The separation of epidermal cells from the old cuticle is known as **Apolysis.**
205. The silk fibre is composed of two proteins fibrion and sericin, accounting for about **75% & 25%.**
206. The silk fibre protein **Fibroin** is synthesized in cells of the posterior silk glands and secreted into the lumen and transported to the middle silk gland for storage.
207. The silk fibrion protein is composed of predominantly three amino acids **Glycine, Alanine and Serine.**
208. The size and weight of silkworm eggs is larger and heavier in **European races.**
209. The smallest haemocyte is **Prohaemocyte.**
210. The substance that inhibits juvenile hormone synthesis by corpora allata is **Allostatic Hormone.**
211. The substance that stimulates juvenile hormone production is **Allotropic Hormone.**
212. The total haemocyte count is higher during **Moulting stage.**
213. The two anterior silk glands, which serve as ducts, converge near the oral cavity, forming a scleroid structure **Spinneret** through which the silk is extruded as fibre.

214. The type of diapause induced by Photoperiod requires several weeks to months of treatment to low temperatures for diapause termination is called **Eudiapause.**

215. The type of diapause terminated by reversal of inducing agents is called **Oligopause.**

216. The type of haemocyte involved in silk production in *Bombyx mori* is **Spherulocyte.**

217. The type of haemocyte not present in silkworm haemolymph is **Leucocyte.**

218. The type of metamorphosis found in silkworms is called **holometabola.**

219. The type of secretion of digestive enzymes where cells do not break down and may discharge their secretion through the striated border is **merocrine.**

220. The universal energy currency of the cell is **ATP.**

221. The waste product produced by the malpighian tubules and emptied into the hindgut is called as **primary urine.**

222. The weight of the silkworm egg is **0.5-0.6 mg.**

223. The **white rot egg** was earlier referred to as the white dead egg and was considered one of the types of dead eggs.

224. The type of parthenogenesis in which individuals of either sex are produced **amphitoky.**

225. The type of parthenogenesis in which only female offspring are produced is **thelytoky.**

226. The type of parthenogenesis in which only male offspring are produced is **arrhenotoky.**

227. Usually, eggs of Chinese and Japanese races are **smaller** than those of the European races.

228. Vasa deferentia receives the sperm from the testes and facilitate their transport to the **ejaculatory duct.**

229. **Vasa deferentia** receives the sperm from the testes and facilitate their transport to the ejaculatory duct.

230. **Vit-C** is principally a water-soluble vitamin and is one of the most indispensable nutrients required by silkworm for normal growth and development.

231. Vitellin is a protein found only in the **females.**
232. What are the accessory organs of excretion in insects are **nephrocytes, urate cells, goblet cells.**
233. Which of the following is a neuromodulator: **Dopamine, Serotonin, Octopamine.**
234. Which of the following is a neurotransmitter is **ACH, GABA.**
235. Which of the following is an excitatory neurotransmitter in insects is **L-glutamate.**
236. Which of the following is the inhibitory neurotransmitter in insects is **GABA.**
237. Which of the following is the reducing disaccharide in silkworm haemolymph is **maltose.**
238. Which of the following is the reducing sugar in silkworm haemolymph is **Glucose.**
239. Diapause hormone structure was first described by **Yamashita.**
240. Yellow needle-like crystals are present in the cells of Malpighian tubules of the fifth instar larva of silkworm known as **lactoflavin and vitamin B2.**
241. Young silkworms of Bombyx mori (chawki worms) prefer **tender Leaves** since higher water and protein contents suit best the fast-growing chawki worms.

4

SILKWORM SEED PRODUCTION

1. Reproductive seeds of mulberry silkworm are multiplying in **Breeding Station.**
2. Commercial seeds of mulberry silkworm are multiplying in **grainages.**
3. Industrial seeds are also known as **commercial seeds.**
4. The system of seed multiplication followed in India is **3-tier system.**
5. Basic seed forms /P3 Stations are called as **temples of seed organisation**.
6. **2% formalin** is used for surface sterilisation of layings.
7. Oviposition temperature is **25°C** and RH **75%.**
8. Light regime of **10 hrs** and **14 hrs** darkness promotes proper emergence.
9. Moth emergence of BV takes **13-14 days.**
10. Moth emergence of MV takes **10-11** days.
11. Cold storage of pupae at **7°C** for 1 or 2 days for Synchronisation.
12. Acid-treated hibernated silkworm eggs are cold Stored for maximum duration of **20 days.**
13. White pluffy eggs occur when silkworm eggs are preserved or transferred from **high temperature to low temperature**.
14. Grainage orientation: **East-west -Tropical, North-South-Temperate**.
15. Multivoltine seeds areas: **Kunigal, Magadi, Huliyurdurga, Santhebmavathur and Hebbur**.
16. Multivoltine Seeds markets: **Kempanahalli, Magadi, Huliyurdurga, Veeragowdandodi, Hebbur, Kadur, Kunigal, Santhemavathur, Solur and Chowdan Kuppe.**
17. Bivoltine seed areas: **Anekal, Sirsi, Hassan, KR-Pet.**
18. Bivoltine seed markets: **Sirsi, Hassan, Bangalore, KR-Pet, Sarjapur.**
19. Tamil Nadu has multivoltine seed area at **Denkanikotti** and **Talavadi.**

20. Andhra Pradesh has multivoltine seed area at **Gorantla** and **Palamaner.**
21. UV and BV can be made to hatch either through **artificial treatment** (hydrochlorization) or **preservation at low temperature**.
22. Embryonic stage ideal for chilling of BV eggs in **stage 5** (24 hrs) to **stage 7** (48 hrs).
23. Short-term chilling of silkworm eggs is cold stored at **5°C** for **30-40** days.
24. In long-term chilling silkworm eggs are cold stored for a maximum duration of **60-70 days.**
25. The hibernated silkworm eggs after releasing from cold storage must be subjected to acid storage within **1-3 hrs.**
26. Acid-treated hibernated silkworm eggs are cold stored for a maximum duration of **20 days.**
27. For acid treatment specific gravity of HCl should be **1.0642 at 46°C.**
28. Duration of acid treatment is related to the breed of silkworm:
29. Japanese: **5-6 min.**
30. Chinese: **4-5 min.**
31. European: **6-7 min**.
32. While treating the hybrids the duration of dipping the eggs is determined as the **Mid value** of the **Duration of parents**.
33. For cold acid treatment the specific gravity of the HCl should be **1.110 at 15°C or 1.105 to 1.119 at room temperature.**
34. Cold acid treatment is suitable for **early-stage** embryos.
35. Hill amelioration stations:
36. **B R hills - Karnataka**
37. **Coonoor - Tamil Nadu**
38. **Kalimpong- West Bengal**
39. For synchronisation female cocoon can be refrigerated on the **7th or 8th day** of spinning at **8°C for 3 days** only and male cocoon for **7 days.**
40. **Hydrometer** is used to find out specific gravity of acids.
41. In male pupa there is a dot in between **8th and 9th abdominal** segments.

42. In female pupa there is a X-like slit on **8th abdominal segment** at ventral side.
43. The intensity of light sufficient to induce moth emergence is **5 Lux.**
44. Moth emergence **6-8 AM.**
45. Cellular rearing is practiced in **P3 station.**
46. Batch rearing is practiced in **P2 station.**
47. Mass rearing is practiced in **P1 station.**
48. Mating periods suggested for production of productive seeds should be less than **120- 150 minutes.**
49. Silk moth lay eggs in **mono-layer or single layer.**
50. Silkworm eggs are transported from one place to another after **second and third days of oviposition.**
51. Mysore silkworm seed act was implemented in Karnataka **1952.**
52. Eggs of **European races** have thicker chorion compared to Japanese and Chinese races.
53. Anterior part of silk gland is **2 cm** long is a thin duck composed of **250 cells.**
54. Middle part of silk gland is **7 cm long** and is consist of **300 secretory cells** producing silk sericin.
55. Posterior part of silk gland is **15 cm** long and consist of **500** secretory cells.
56. **National grid** is distribution system of silkworm egg.
57. **Winnowing** of unfertilized eggs is completely practised during loose egg production.
58. During oviposition, **cellule** is used to create semi dark condition.
59. Grainage activity lasts for about **20 days.**
60. During silkworm egg development, lead pigmentation stage is also known as **pin head stage.**
61. During silkworm egg development, blue egg stage is also known as **body pigmentation.**

62. During silkworm embryonic development, the largest embryonic stage is **C_2 stage.**
63. During silkworm embryonic development process, in F_4 stage eye spots appears **F_3 stage trachea.**
64. If developing eggs of *B. mori* are exposed to long photoperiod, the resultant female produces high proportion of **diapause eggs.**
65. If developing eggs of *B. mori* are exposed to short photoperiod, the resulting adult female produce high proportion of **non-diapausing eggs.**
66. The production of 100 % true hybrid is ensured by doing sex separation at **Pupal stage.**
67. During mother mouth examination **0.6% of K_2CO_3** solution is used.
68. During 'test hatching' test **0.25 grams** of silkworm eggs samples are collected from each lot.
69. The hierarchy of basic seed organisation is in the order of **P3 → P2 → P1.**
70. The protein content in the raw fresh pupa of *B.mori* is **79%.**
71. JH is secreted by **corpora allata.**
72. Moulting hormone secreted by **prothoracic gland.**
73. Diapausing hormone is secreted by **sub-oesophagus ganglion.**
74. The **3- hydroxy kynurenine** is present in diapausing eggs.
75. When univoltine eggs are exposed to low temperature and short-day length it leads to **diapausing eggs.**
76. In diapausing silkworm egg, the activity of **esterase 'A' enzyme** is inhibited.
77. After depairing the female silkworm is allowed to **24 hours** for oviposition.
78. Male and female cocoons refrigerated to maximum of **7 and 3 days** respectively.
79. Bivoltine virgin female moths are preserved at **5° C** for **3 and 4-5 days** in summer and rainy season respectively.
80. The newly laid silkworm egg, glycogen accounts to **2.5%.**
81. During silkworm egg development neurohormones triggers the synthesis and synthesis and release of **ecdysone** from the ovary.

82. The micropylar canal on silkworm egg surface is normally **3-4** in number.
83. The first maturation division is silkworm egg terminates **5-6 minutes** after position.
84. In silkworm egg, second maturation division take place in **60 minutes** after oviposition.
85. In silkworm egg, second maturation division completes in about **20 minutes.**
86. In silkworm egg, the reduction division takes approximately **90-100 minutes** after oviposition.
87. In **stage-4** of the silkworm embryo, the germband formation takes place.
88. In **stage-5** cephalic lobe and caudal lobe formation takes place in silkworm embryo development.
89. The embryonic **stage-5 to stage-7** is ideal for chilling.
90. The embryonic **stage-8** of silkworm egg ideal for different hibernation schedule.
91. **Stage-15** of embryo is suitable for refrigeration at a 2.5°C temperature for 10-40 days.
92. In silkworm egg stage, metabolic energy is derived from oxidation of **triglycerol.**
93. In silkworm egg, **phospholipids** are a major component of yolk and act as a precursor for the formation of embryonic cells.
94. Lipid concentration is **higher** in diapausing silkworm than in non-diapausing.
95. **P_1 DFL's** are considered as input for rising quality seed cocoons through the seed farmers modified seed areas.
96. Generally obligatory diapause occurs in **univoltine** silkworm.
97. Facultative diapause occurs in **bivoltine** silkworm.
98. In Japanese silkworm rearing, the fecundity is **600-700 eggs/moth.**
99. P1 seed farmers are also called **registered multivoltine seed cocoon rearers.**
100. **Bilidevalaya** is considered a P4 silk farm for multivoltine races.

5

SILKWORM REARING

1. Silkworm belongs to the family **Bombycidae.**
2. The scientific name of domesticated mulberry silkworm is ***Bombyx mori* L.**
3. Ancestor of *Bombyx mori*: ***Bombyx mandarina***
4. The stages in the life cycle of silkworm is **4 stages.**
5. The period between successive moults is called **instar.**
6. The reproductive stage of silkworms is **adults.**
7. Diapausing eggs are called as **Kuradone eggs.**
8. Non-Diapausing eggs are called as **Nomodone eggs.**
9. Diapausing eggs are produced by **univoltine and bivoltine silkworms.**
10. Non-diapausing eggs are produced by **multivoltine silkworms.**
11. Classification of silkworm based on voltinism: **Univoltine, Bivoltine, Multivoltine.**
12. Silkworms that complete one generation per year are called **univoltine.**
13. Silkworms that complete two generations per year are called **bivoltine.**
14. Silkworms that complete three generations per year are called **multivoltine.**
15. Classification of silkworm based on moultinism: **Trimoulters, Tetramoulters, Pentamoulters.**
16. Silkworm that moults three times during the larval period is called **trimoulters.**
17. Silkworm that moults four times during the larval period is called **tetramoulters.**
18. Silkworm that moults five times during the larval period is called **pentamoulters.**

19. Amylase activity is high in **multivoltine.**
20. European races are **univoltine.**
21. Chinese and Japanese races are **univoltine and bivoltine.**
22. Tropical races are **polyvoltine/ multivoltine.**
23. The type of head silkworm larva has is **hypognathus.**
24. The type of mouth silkworm larva has is **biting and chewing type.**
25. Silk moth antennal type is **bipectinate.**
26. The number of spiracles present in silkworm larva are **9 pairs.**
27. Silkworm pupa is **obtect.**
28. The most popular method of rearing in India is **shelf-rearing (shoot rearing) or tray-rearing.**
29. The moulting period for I instar larva is **20 hours.**
30. The moulting period for II instar larva is **20 hours.**
31. The moulting period for III instar larva is **24 hours.**
32. The moulting period for IV instar larva is **30 hours.**
33. Rearing trays have a diameter of **138 cm.**
34. Use of feathers in silkworm rearing - **spread the young worms/ brushing.**
35. The process of preserving silkworm eggs under optimum temperature, RH, photoperiod and air current to facilitate proper development of the embryo and ensure uniform hatching called **incubation.**
36. Optimum temperature for egg incubation: **25±1°C.**
37. Optimum relative humidity for egg incubation: **70-80%.**
38. Optimum photoperiod for egg incubation: **16 hours of light and 8 hours of darkness.**
39. Optimum air current for incubation: **0.3 m/s.**
40. Transportation of silkworm eggs done during **cooler hours of the day or hibernation period (post or pre-hibernation).**
41. Silkworm eggs are transported using **egg carrying box.**
42. The layer of silkworm egg acting as a barrier to water loss is **endochorion.**

43. The duration of egg incubation is **10 days** (or from egg laying to hatching).
44. Methods of egg incubation- **constant temperature incubation and raised temperature incubation.**
45. The method of incubation used for incubating non-hibernating and acid-treated eggs is **Constant temperature incubation.**
46. The method of incubation used for incubating hibernating eggs is **Raised temperature incubation.**
47. Before 48 hours of hatching, a black spot appears on eggs, this condition is referred to as **head pigmentation stage or pin head stage.**
48. Eggs turn a bluish colour one day before hatching this is referred to as **body pigmentation stage or blue head stage.**
49. Silkworm eggs in the body pigmentation stage are artificially confined to dark phase prior to hatching and the process is referred to as **black boxing.**
50. The process of transferring newly hatched larva from the egg sheet onto the rearing sheet is called **brushing.**
51. Newly hatched larvae are called **ants or crawlers.**
52. Hatching of silkworm larvae starts at **5-6 AM.**
53. The best time for brushing is **10 AM (morning hours).**
54. Methods of brushing - **loose eggs brushing and egg card brushing.**
55. Types of brushing of ants from egg card - **tapping method, feather method, husk method, net feeding method.**
56. The optimum temperature required for rearing of I Instar silkworm larva is **26-28°C.**
57. The optimum temperature required for rearing of II Instar silkworm larva is **26-28°C.**
58. The optimum temperature required for rearing of III Instar silkworm larva is **25-26°C.**
59. The optimum temperature required for rearing of IV Instar silkworm larva is **24-26°C.**
60. The optimum temperature required for rearing of V Instar silkworm larva is **24-25°C.**

61. The optimum relative humidity required for rearing of I Instar silkworm larva is **85%.**
62. The optimum relative humidity required for rearing of II Instar silkworm larva is **85%.**
63. The optimum relative humidity required for rearing of III Instar silkworm larva is **80%.**
64. The optimum relative humidity required for rearing of IV Instar silkworm larva is **75%.**
65. The optimum relative humidity required for rearing of V Instar silkworm larva is **70%.**
66. The length of newly hatched silkworm larvae is **3 mm.**
67. The newly hatched silkworm larvae are commonly called **ants or kengo.**
68. The young-age silkworm rearing is called **chawki rearing.**
69. Chawki worms require **80% moisture, 27% protein and 11% carbohydrates.**
70. In early instars, ingestion of mulberry leaves is **slow** but digestion is **high.**
71. Chawki rearing usually lasts for **7-8 days.**
72. Methods of chawki rearing - **box rearing, tray rearing.**
73. The box rearing method is practised when **temperature and humidity are lower.**
74. Rearing of grown-up silkworms *i.e.*, 3rd, 4th and 5th stage to spinning is called **late-age rearing.**
75. The amount of mulberry ingestion is more and digestion is less in **late-age silkworms.**
76. Methods of late-age silkworm rearing: **Shelf rearing, tray rearing, floor rearing and shoot rearing.**
77. The sparsely feeding stage means - **when silkworm larva feeds less quantity.**
78. Larva stops feeding during the **moulting stage.**
79. The recommended frequency of feeding in shoot rearing is **three feeds per day.**

80. The recommended frequency of feeding in tray rearing is **four feeds per day.**
81. The silkworm rearing stands are made up of **locally available wood or bamboo or PVC pipes.**
82. The height of silkworm rearing stand is **2.5 m.**
83. The use of paraffin paper in silkworm rearing is to **prevent leaf moisture evaporation and maintain humidity.**
84. The use of foam rubber strips in silkworm rearing is to **maintain humidity.**
85. The use of chopsticks in silkworm rearing is to **pick diseased worms from rearing bed.**
86. The number of bed cleanings recommended in case of shoot rearing is **once** (after fourth moult).
87. The ideal size of chopped leaf for I instar silkworm larva is **0.5-2 cm.**
88. Ideal size of chopped leaf for II instar silkworm larva is **2-4 cm.**
89. Ideal size of chopped leaf for III instar silkworm larva is **4-6 cm.**
90. The amount of ingestion and digestion per given time **increases** with an **increase in temperature.**
91. High humidity **shortens larval duration** by accelerating the physiological junction of silkworm
92. Low humidity **increases the growing period.**
93. In general, pure lines and multivoltines require **lower temperatures** than bivoltines at the same stage.
94. The appliances needed for maintaining optimum rearing conditions are **thermometer, hygrometer, humidifier.**
95. The rearing house should be **north-facing.**
96. The direction of windows in late-age silkworm rearing: **North and South.**
97. Accessary rooms/ ante-rooms of rearing house - **mulberry leaf storage room, mounting room.**
98. Herold's gland is present in **male silkworm larvae** (ventral median line between the 8th and 9th abdominal segments).

99. Ishiwata's gland is present in **female silkworm larvae.**
100. Ishiwata's fore gland is located on the **8th abdominal segment.**
101. Ishiwata's hind gland is located on the **9th abdominal segment.**
102. The identification mark of the female silkworm pupa is **vertical line ('X' mark) in the centre of the 8th abdominal segment.**
103. The identification mark of male silkworm pupa is **dark spot on the 9th segment.**
104. The silk moths lost their flight **due to several years of domestication.**
105. The appliance used to support spinning larvae is **mountage.**
106. The process of removing litter periodically from the bed is called **bed cleaning.**
107. Methods of bed cleaning - **husk method, net method, combined net and husk method, rope method.**
108. Moulting is also called as **ecdysis.**
109. The periodical process of shedding old cuticle and the formation of new cuticle is known as **moulting.**
110. The cuticle part discarded during moulting is **exuvia.**
111. The time interval between two subsequent moults is called **stadium.**
112. Moulting is controlled by **endocrine glands (prothoracic gland).**
113. The process of transferring ripe worms to mountages is called **mounting.**
114. The final instar larva when stops feeding and is ready to spin a cocoon is called **ripe worms.**
115. The ripening stage is attained for multivoltine and bivoltine races in tropical conditions on **6-7 days of feeding in 5th instar**
116. The destruction of disease-causing pathogens is called **disinfection.**
117. The classification of disinfectants - **room disinfectants, bed disinfectants and general disinfectants.**
118. Examples of room disinfectants - **5% bleaching powder, Sanitech, Serichlor.**
119. Examples of bed disinfectants - **Amruth, Vijetha, Suraksha.**

120. The eco-friendly, plant-based formulation for suppressing grasserie and flacherie is **Amruth.**
121. The disinfectant used for effective sanitation is **Serifit.**
122. The hormonal formulation developed by CSRTI, Mysuru for attaining uniform maturity in silkworms is **'Sampoorna'**.
123. The recommended dose of 'Sampoorna' application - **4 litres of diluted hormones for 100 DFLs.**
124. Urine excreted by ripe worms is called as **meconium.**
125. Before spinning, nearly **40 ml of urine (meconium)** is excreted by 100 worms.
126. Methods of mounting - **hand picking, net method.**
127. Types of mountages - **Chandrike (bamboo mountage), Rotary mountage, Plastic collapsible Mountage and Bottle brush mountage.**
128. Humidity in the mountage room is optimum for two important reasons:
129. **Firstly**, mounted larvae lose as much as 40% of their body weight due to natural evaporation.
130. **Secondly**, the larvae tend to discharge liquid excreta before the commencement of spinning.
131. The ideal relative humidity for cocoon spinning is **60-70%.**
132. The ideal temperature for cocoon spinning is **24°C.**
133. The ideal air current speed for cocoon spinning is **1 m/s.**
134. The cocoon spinning process varies at the rate of **0.4-1.5 cm/s.**
135. The movement of the silkworm head during spinning *i.e.*, the baves deposited in series of short waves that looks like **number '8' or '∞'.**
136. Larva moves about **70-80 times/ min** in the process of spinning.
137. Under optimum temperature and humidity, the spinning of the cocoon is completed in **48 to 60 hrs.**
138. The outer floss and the innermost pelade together made the waste **frisons.**
139. The outer layer of the cocoon, floss is also called as **blaze.**
140. The cocoon and the bave or filament are formed at the rate of **15cm/min.**

141. Cocoon harvesting is done on the **5th or 6th** day of spinning.
142. The fibre from two silk glands comes out through spinneret independently called **brins.**
143. The process of removing the flossy layer from the cocoon is called **deflossing.**
144. Removing odd-shaped and defective cocoons from the lot is called **sorting.**
145. Deformed cocoons may form due to **improper mounting frames.**
146. Mountage-pressed cocoons are also called **scaffold-pressed cocoons.**
147. The cocoons stained inside are called **dead cocoons or melted cocoons.**
148. The cocoon spun by two worms producing a filament that does not unwind smoothly and tangles easily is called **double cocoon.**
149. The only carbohydrate found in the digestive fluid of silkworms is **amylase.**
150. Mention 10 essential amino acids required for silkworms:
151. **Arginine, Histidine, Leucine, Isoleucine, Lysine, Methionine, Phenylalanine, Threonine, Tryptophan, Valine.**
152. Silkworms vomit their intestinal juices at **12 to 13% CO_2** in the air and die at a **15%** concentration.
153. The safe limit for the gases in the late stage of silkworm rearing is **carbon dioxide - 1-2%, formaldehyde - 1%, SO_2 - 0.02% and ammonia - 0.1%.**
154. **Seriphene** is a broad-spectrum general disinfectant developed by KSSRDI, Bengaluru.
155. **Sericillin** is a mixture of Lime + Bleaching powder + Fungicide.
156. In shoot rearing method, labour is reduced by **60%** in 4th and **50%** in 5th instar.
157. In shoot rearing method, leaf saved is **25%** in 4th and **10%** in 5th instar.
158. Silkworms consume **14% and 80%** of the total quantity of leaves required in 4th and 5th instar, respectively.
159. Botanical-based bed disinfectants are **Ankush, Vijetha Forte and Vijetha Green.**

160. **Sanjevini** is also called 'A' powder.
161. **Suraksha** is also called 'B' powder.
162. **Poshak** to Source of nutrients for feed supplementation to silkworms during 5th instar.
163. **Serimore** is a plant-based extract used as silkworm growth promoter.
164. **Chethana** is a phyto-ecdysone hormone that induces spinning in silkworms developed by KSSRDI.
165. **Samrudhi** - JH spray increases cocoon yield by 8-10kg/100 DFLs.
166. **C-Nichi** was originally a Japanese bivoltine introduced to India in Mysore. Due to continuous rearing, it degenerated into a multivoltine race.
167. **Nistari** was introduced in 1881 from China and continues to dominate commercial rearing in entire north India, especially West Bengal.

6

PESTS AND DISEASES OF SILKWORM

1. The tachinid that infests and breeds on the silkworm is **uzi fly**.
2. Indian uzi fly – ***Exorista bombycis* (Louis)**.
3. Tasar uzi fly – ***Blepharipa zebina* (Walker)**.
4. Japanese uzi fly – ***Crossocosmia sericariae* (Rondani)**.
5. Hime uzi fly – ***Ctenophorocera pavida* (Meigen)**.
6. The egg laying preference of uzi flies on silkworm – **dorsal intersegmental regions**.
7. Silkworm more prone to uzi fly incidence is **late age silkworms**.
8. Uzi fly eggs hatches **61-62 hours after egg laying**.
9. The sign of uzi fly infestation on the silkworm body is appearance of **black scar**.
10. Uzi maggot undergoes **three instars**.
11. Uzi maggot is **fusiform**.
12. The larval period of the uzi fly is **4-7 days**.
13. The mature uzi maggot crawls out of the host and transforms into pupa in **6-8 hours**.
14. The pupal duration of uzi fly is **7-18 days**.
15. The characteristic pattern observed on adult uzi fly is **four longitudinal black lines on the thoracic region**.
16. The antenna of uzi fly – **aristate antenna**.
17. The host preference of uzi fly – ***Bombyx mori* > *Samia cynthia ricini* > *Antheraea mylitta* > *A. assamensis***.
18. The most common method followed to exclude the uzi fly from the silkworm is using **nylon nets/mesh**.
19. Quarantine measure to be employed to prevent uzi fly introduction to new areas is **transporting silkworm seeds instead of seed cocoons**.

20. The combination product, uzicide developed by CSRTI, Mysuru is an effective **ovicide/ oviposition deterrent**.
21. The use of uzi traps/tablets against uzi flies is a **physical** control.
22. The chemical known to cause sterility in adult uzi flies is **diflubenzuron**.
23. Recommended parasitoids against uzi fly are ***Nesolynx thymus*** (2 pouches/100 DFLs) and ***Dirhinus anthracia***.
24. *Nesolynx thymus* is a **pupal parasitoid**.
25. Uzi powder is an ovicidal formulation developed by **KSSRDI, Bengaluru**.
26. The hymenopteran pest known to attack silkworms on rearing beds and infest the stored cocoons is **fire ant (*Solenopsis geminata*)**.
27. The cosmopolitan insect pest known to attack silk moths in grainage is **earwig (*Labia arachidis*)**.
28. The insect pest known to cause damage to stored silk cocoons, raw silk yarn and silk fabrics is **Dermestid beetle** (*Dermestes* sp. and *Anthrenus* sp.)
29. The characteristic circular holes on the cocoons are the sign of infestation of **carpet beetle (*Attagenus fasciatus*)**.
30. The distinguishing feature of cabinet beetle, *Anthrenus verbasi* from other members of dermestid beetle is **adult's presence on plants and preference to daylight**.
31. Recommended chemical in the management of dermestid beetle is **0.028% Deltamethrin** (spraying at an interval of 6 months).
32. The non-insect pest that develops black specks on the silkworm body, bead-like excreta attached to anus, yellowish-green fluid vomiting and if severely parasitized, excretes black fluid and starts putrefying is **straw itch mite (*Pyemotus ventricosus*)**.
33. The pathogenic factor that leads to differences in genetic makeup, developmental stage and sex of the silkworm are **constitutional factors**.
34. Principles of silkworm disease control – **prevention, diagnosis and increasing vigour**.
35. An impairment in the normal state of the living organism due to a specific pathophysiological response is called **disease**.

36. A group of symptoms that disrupt normal body functions but does not have a known cause is called **disorder**.
37. The determination of a disease from its signs, symptoms, etiology, pathogenesis, physiopathology and morpho pathology is **diagnosis**.
38. The study of the development of a disease in relation to the environment is called **epizootiology**.
39. The outbreak of a disease that affects a huge population of a species at the same time is called **epizootic** ('epidemic' in case of the human population).
40. The branch of immunology concerned with the *in-vitro* antigen-antibody reactions is **serology**.
41. The substance that stimulates the formation of a specific antibody when introduced is **antigen**.
42. The substance that enhances the antigenicity of another substance is **adjuvant**.
43. The gamma globulins produced in the host system in response to the introduction of foreign substance is **antibody**.
44. The process in which the antibodies are directed against the surface of antigens in a host system so as to form the aggregates is called **agglutination**.
45. The solid, semisolid or liquid preparations employed for the cultivation and identification of microorganisms is **culture media**.
46. Agar-agar is preferred over gelatin as a solidifying agent because it has a **higher melting point** (85°to 100°C, while gelatin melts at 35°C).
47. The process of freeing any object or substance from all forms of life is called **sterilization**.
48. The process of selective destruction of heat-sensitive microbial populations through the application of mild heat is called **pasteurization**.
49. The bright field microscopes are **simple microscope** and **compound microscope**.
50. A light microscope equipped with a special condenser and objective to brightly illuminate the microorganisms in the specimen against a dark background is **dark field microscope**.

51. A microscope to observe the unstained live microorganisms clearly in the presence of both annular and a phase-shifting element is **phase contrast microscope**.
52. The microscope where the electron beam passes directly through the specimen and is viewed on the screen is **transmission electron microscope (TEM)**.
53. The microscope where the electron beam is reflected from the surface of the specimen and its surface topography is observed in minute details on a screen is **scanning electron microscope (SEM)**.
54. The measurement of minute objects using a microscope is called **micrometry**.
55. A practical and common laboratory technique used in enumerating the live microorganisms in any given material is **dilution plate technique**.
56. Turbidometric method of microorganism enumeration is employed for **bacterial cultures**.
57. If all the cells in the culture sample have a common origin and are simply descendants of the same cells, it is called **pure culture**.
58. Gram-positive bacterial cells on gram staining appear **violet**.
59. Gram-negative bacterial cells on gram staining appear **red/pink**.
60. In endospore staining, the endospores appear **green** and the other cell parts and vegetative cells appear **red**.
61. The staining used to differentiate the NPV and CPV polyhedral inclusion bodies is **methyl green pyronin-G staining** (NPV – blue; CPV – red).
62. Pebrine, a microsporidian disease in silkworm (*Bombyx mori* L.) is caused by ***Nosema bombycis* (Naegeli, 1857)**.
63. Pebrine disease was first recorded in **France (1845)**.
64. In 1849, Guerin-Meneville from France named the pebrine pathogen as **hematozoid**.
65. Microsporidia was characterised based on the production of minute oval spores by **Leybig and Balbiani** (1850-1870).
66. The name Pebrine was coined by **De Quadrefagues (1860)**.
67. Transovarial transmission of pebrine was discovered and mother moth examination for diagnosis was suggested by **Louis Pasteur (1865-1870)**.

68. Balbiani's book that revealed that pebrine causing pathogen belongs to Microsporidia under Sporozoa of Protozoa is "**Lecons Surles Sporozoairls**" (1882).
69. Various stages in the life cycle of *Nosema bombycis* was described by **Stempell** (1909).
70. The characteristic symptom of pebrine disease in the advanced stage of infection is **pepper-like spots on the diseased larvae**.
71. In Bengali, pebrine disease is known as "**Kata**" or "**Matha Kata**".
72. In Kannada, pebrine disease is known as "**Gantu**" or "**Gantu Roga**".
73. The mode of transmission of pebrine disease are **per-oral, trans-ovarial and trans-ovum**.
74. In temperate countries, pebrine disease is common during **summer**.
75. In tropical countries, pebrine disease is common during **winter**.
76. The polar filament of *Nosema bombycis* has 12 coils with an angle of tilt of atleast **49 degrees**.
77. The dimorphic development of *Nosema bombycis* spores in the larvae of *Bombyx mori* occurs in **midgut epithelium**.
78. The *Nosema* infected silkworms undergo super-numerary moults or extension of larval period due to the presence of **juvenile hormone (JH)**.
79. The two stages in the life cycle of *Nosema bombycis* are **spore stage and vegetative stage**.
80. The extra-cellular phase in the vegetative stage of *Nosema bombycis* infection that is motile, multiplies by binary fission and responsible for autoinfection of fresh cells in the same host is called **planont**.
81. The intra-cellular phase in the vegetative stage of *Nosema bombycis* infection that is immotile and produces spores that can infect new hosts is called **meront**.
82. The *Nosema bombycis* spore germination happens at the midgut **distal end**.
83. Primary site of *Nosema bombycis* infection is **midgut**.
84. Secondary site of *Nosema bombycis* infection is **malphigian tubules** (followed by fat bodies, muscle fibres, silk glands and gonads).

85. The symptoms of *Nosema bombycis* infection on silkworm eggs – **laid in clumps, lack of uniformity in shape, reduction in size, poor attachment, pale colouration, high percentage of unfertilised and dead eggs, decrease in hatching percentage**.

86. The major symptom of *Nosema bombycis* infection on early instars silkworms – **no spacing of body hairs**.

87. The major symptoms of *Nosema bombycis* infection on late age silkworms – **unequal larval size, wrinkled skin with rusty colour, black pepper-like irregular spots appear on the body, mostly on the caudal horn and lateral to prolegs, gut becomes opaque and silk glands shows white pustules or spots**.

88. The major symptoms of *Nosema bombycis* infection on silkworm pupa – **cocoon spun is flimsy, malformed, pale coloured with thin ends; pupa is swollen, lustreless and irregular black spots appear on the body wall**.

89. The major symptoms of *Nosema bombycis* infection on silk moth – **delayed moth emergence, deformed antennae, clubbed wings with vesicles or black spots, naked moths without abdominal scales, low fecundity and irregular egg laying**.

90. The order of resistance to pebrine disease – **Chinese breeds > Japanese breeds > Indian breeds > European breeds**.

91. The order of resistance to pebrine disease – **Multivoltine > Bivoltine > Univoltine**.

92. The most commonly used pebrine detection technique is **Mother Moth Examination (MME)**.

93. The chemical reagents or solution used in Mother Moth Examination are **K_2CO_3 and KOH**.

94. The occluded viruses infecting silkworms are **Baculovirus** and **Reovirus**.

95. The non-occluded viruses infecting silkworms are **Picornavirus** and **Parvovirus**.

96. The antiviral activity of the digestive juice of silkworm larvae is due to the presence of **Red Fluorescent Protein (RFP)**.

97. The nuclear polyhedrosis virus (NPV), a Baculovirus causes **grasserie**.

98. The nucleic acid type found in the nucleocapsids of *Bm*NPV is **double stranded (ds) DNA**.
99. The molecular weight of ds DNA in *Bm*NPV is **50×10^6 daltons**.
100. In Karnataka, grasserie disease of *Bombyx mori* is called "**Halu Hula**".
101. In West Bengal, grasserie disease of *Bombyx mori* is called "**Rasa**".
102. In Andhra Pradesh, grasserie disease of *Bombyx mori* is called "**Palu Purugu**".
103. In Tamil Nadu, grasserie disease of *Bombyx mori* is called "**Pal Poochi**".
104. In *Bombyx mori*, the NPV polyhedra is of **dodecahedral** shape.
105. The virions in the silk gland cells are of **MNPV** type (M = multiples, *i.e.*, aggregation of nucleocapsids).
106. The *Bm*NPV virion nucleocapsid is **rod shaped**.
107. Primary site of infection of occluded virus is **midgut epithelial cells**.
108. The *Bm*NPV infection in silkworms is common in **5th instar larvae**.
109. The *Bm*NPV infection in silkworms occur in **all seasons**.
110. The increase in temperature and decrease in humidity has **direct** effect on *Bm*NPV development (↑ temperature + ↓ humidity = ↑ *Bm*NPV development).
111. The major symptoms *Bm*NPV infection in later stages – **shiny exoskeleton, intersegmental swelling, larvae fail to moult and move aimlessly, exoskeleton ruptures and turbid white haemolymph oozes out**.
112. The cytoplasmic polyhedrosis virus (CPV), a Reovirus causes **flacherie**.
113. The silkworm cytoplasmic polyhedrosis was first described by **Ishimori (1934)**.
114. The nucleic acid type found in the *Bm*CPV is **double stranded (ds) RNA**.
115. The molecular weight of ds RNA in *Bm*CPV is **35×10^6 daltons**.
116. The *Bm*CPV polyhedra are of **hexagonal and tetragonal** shape.
117. The electron microscopic study of the histopathology of *Bm*CPV was made by **Tsujita (1955)**.

118. Primary site of *Bm*CPV infection is **posterior end of midgut**.
119. *Bm*CPV multiplies mainly in the cytoplasm of **columnar cells of midgut epithelium**.
120. *Bm*CPV infection is common in **silkworm larvae**.
121. The order of susceptibility of *Bm*CPV in silkworms is **early instars > late instars**.
122. *Bm*CPV incidence is prevalent during **summer**.
123. Predisposing environmental factors for *Bm*CPV infection are – **fluctuating temperature and high humidity**.
124. The major symptoms *Bm*CPV infection in later stages – **larvae become soft, flaccid, inactive, vomits gut juice and faeces become soft with high moisture content, sometimes chain-like excreta are observed**.
125. The densonucleosis virus (DNV), a Parvovirus causes **densonucleosis**.
126. The nucleic acid type found in the *Bm*DNV is **single stranded (ss) DNA**.
127. The molecular weight of ss DNA in *Bm*DNV is **6×10^6 daltons**.
128. The densonucleosis virus was first isolated from **greater wax moth larvae, *Galleria mellonella* (1964)**.
129. The Kenchu disease, a prevalent flacherie in Karnataka is attributed to ***Bm*DNV-2**.
130. The *Bm*DNV was first isolated by **Shimizu (1975)**.
131. The *Bm*DNV polyhedra is an **icosahedron** (20 faced polyhedra).
132. Primary site of *Bm*DNV infection is **midgut epithelium**.
133. *Bm*DNV infection is common in **early instar silkworms**.
134. Predisposing factor for *Bm*DNV infection is **high temperature**.
135. The major symptoms *Bm*DNV infection are – **head and thorax upheld, motionless, flaccid body, irregular shaped faeces or diarrhoea and gut is filled with brownish fluid**.
136. The infectious flacherie virus (IFV), a Picornavirus causes **infectious flacherie**.
137. The nucleic acid type found in the *Bm*IFV is **single stranded (ss) RNA**.

138. The molecular weight of ss RNA in *Bm*IFV is **2.5×10^6 daltons**.

139. The infectious flacherie disease in *Bombyx mori* was opined to be caused by a virus by **Yamazaki (1960)**.

140. The *Bm*IFV polyhedra is **hexagonal** in shape.

141. Highly susceptible silkworm instar to *Bm*IFV infection is **second instar larvae**.

142. *Bm*IFV multiplies in the cytoplasm of **goblet cells** of midgut epithelium.

143. The major symptoms *Bm*IFV infection are – **shrinkage of body, head and thorax upheld, motionless, flaccid, chain or string like excreta, rectal protrusion, exoskeleton becomes pale with light reddish tinge and variation in moulting duration**.

144. The bacteria associated with higher virulence of *Bm*IFV infection are ***Streptococcus faecalis*** (endogenous) and ***Serratia marcescens*** (exogenous).

145. The flaccidity in the silkworm caused by bacteria is called **bacterial flacherie**.

146. The chronic bacterial disease (or flacherie) of the silkworm digestive tract caused due to multiplication of bacteria in the gut is **bacteremia**.

147. The bacteria associated with bacteremia in silkworm are ***Streptococcus faecalis* and *Streptococcus faecium***.

148. The bacterial infections of **lowers** the gut pH (for example, 8.0 to 9.5).

149. Septicemia in silkworms is most commonly caused by ***Serratia marcescens* and *Bacillus* sp.**

150. Septicemia leading to development of dark brown-reddish tinge spots on dead silkworms is caused by ***Serratia marcescens***.

151. "Rangi" disease in the early instars of silkworms is caused by ***Serratia marcescens***.

152. Septicemia leading to development of dark greenish tinge on the dead silkworms is caused by ***Bacillus* sp.**

153. *Serratia marcescens* in the digestive tract of silkworms is **non-pathogenic** (pathogenic only when they enter the haemocoel).

154. Bacterial toxicosis is caused by ***Bacillus thuringiensis***.

155. "Sotto" disease is caused by ***Bacillus thuringiensis* var. *sotto*** (Ishiwata, 1902).
156. Toxins isolated from *Bacillus thuringiensis* are **three exotoxins (α, β and γ) and one endotoxin (δ)**.
157. The acute bacterial toxicosis in silkworms is caused by the ingestion of **δ-endotoxin**.
158. The major symptoms of acute bacterial toxicosis are - **sudden cessation of feeding, lifting of head, spasms and tremors, larvae wriggles in severe pain and death; corpse is hard to touch with hook-shaped appearance, it turns black and exudes rotten foul-smelling brownish fluid**.
159. The chronic bacterial toxicosis in silkworms is caused by the ingestion of ***Bt* crystal toxin**.
160. The major symptoms of chronic bacterial toxicosis are – **reduced feeding, irregular shaped faeces, occasional vomiting, muscle paralysis, motionless, lose clasping power of legs, flaccidity and death**.
161. The gram-positive rod-shaped motile bacteria with endospores are ***Bacillus* sp**.
162. The gram-positive coccus-shaped non-motile bacteria without endospores are ***Streptococcus* sp**.
163. The gram-negative tiny rod-shaped motile bacteria without endospores are ***Serratia* sp**.
164. Predisposing factors for bacterial diseases are **high temperature and high humidity**.
165. "Thatte roga" in Karnataka is caused due to the synergistic association of ***Streptococcus* sp., *Bacillus* sp. and *Bm*IFV**.
166. The first entomopathogenic fungus observed was ***Beauveria bassiana*** (Bassi, 1835).
167. The white muscardine in silkworms is caused by ***Beauveria bassiana***.
168. In Karnataka, white muscardine on silkworms is called "**Sunnakaddi**" or "**Sunnakattu roga**".
169. In West Bengal, white muscardine on silkworms is called "**Chenaketu**" or "**Chitti**".

170. The hyphal description of *Beauveria bassiana* – **hyaline, smooth, globose to ellipsoidal and appears colourless but light green under microscope**.

171. Primary source of muscardine infection is **conidia**.

172. The duration of one complete life cycle of muscardine fungus on silkworm, *Bombyx mori* in tropical conditions is **4 days**.

173. The duration of one complete life cycle of muscardine fungus on silkworm, *Bombyx mori* in temperate conditions is **10 days**.

174. The symptoms of white muscardine infection on silkworm egg – **infected eggs turn pale or pink, shrunken chorion, eggs fail to hatch**.

175. The symptoms of white muscardine infection on silkworm larvae – **sluggish, inactive, increased pulsation, pale colour, lose appetite, intersegmental and spiracle areas turn brown to black specks, specks ooze oily substance, body turns soft and death occurs; white efflorescence covers the cadaver**.

176. The symptoms of white muscardine infection on silkworm pupa – **infected cocoons appear brighter, lighter in weight and on shaking sounds like dried groundnut pods (mummified cocoons)**.

177. The symptoms of white muscardine infection on silk moth – **sluggish, slightly deformed wings, after death white efflorescence is formed**.

178. Muscardine is prevalent during **monsoon and winter** seasons.

179. Predisposing factors for muscardine disease – **low temperature and high humidity**.

180. The green muscardine in silkworms is caused by ***Metarrhizium anisopliae***.

181. The hyphal description of *Metarrhizium anisopliae* – **delicate, hyaline, smooth, septate, irregularly branched, cylindrical and appears green under microscope**.

182. The symptoms of green muscardine infection on silkworm – **large moist specks with blackish circumference on the larvae, cadaver looks white but turns green and gradually to black; cadaver gets covered with bright greenish conidia**.

183. Japanese green muscardine – ***Spicaria prasina***.

184. Yellow muscardine – ***Paecilomyces farinosus***.
185. Red muscardine – ***Sporosporella uvella***.
186. Orange muscardine – ***Sterigmatocystis japonica***.
187. Hirsutella muscardine – ***Hirsutella satumaensis***.
188. Creamy white muscardine - ***Aphanocladium album***.
189. Aspergillosis is a fungal disease in silkworms caused by ***Aspergillus* sp.** (*A. flavus* and *A. tammari*).
190. Aspergillosis is also called as "**Koji Kabi**" (Japan) or **brown muscardine**.
191. The primary source of aspergillosis infection is **conidia**.
192. The conidial description of *Aspergillus* sp. – **Unicellular, uninucleate, globose/sub globose/oval to elliptical**.
193. Aspergillosis incidence is observed in **young age silkworms** or **chawki stage**.
194. Aspergillosis is prevalent during **monsoon and winter** seasons.
195. Predisposing factors for aspergillosis – **high temperature and high humidity**.

7

SILK REELING TECHNOLOGY

1. The soluble protein present in silk is **sericin.**
2. The insoluble protein present in silk is **fibroin.**
3. The per cent fibroin content in the cocoon shell is **70-80%.**
4. The per cent sericin content in the cocoon shell is **20-25%.**
5. Brin consists of **1 fibroin layer + 3 sericin layer.**
6. The core of silk bave consists of **fibroin.**
7. **Bave** is not continuous.
8. Glycine, alanine and serine amino acids present in the fibroin chain are in the ratio **3:2:1.**
9. Posterior part of silk gland secretes silk substance called as **fibroin.**
10. Middle part of silk gland secretes gummy substance called **sericin.**
11. The main objective of cocoon sorting is to remove **defective cocoons**.
12. Pierced cocoons are mainly due to **tachinid flies (uzi fly).**
13. Japanese breeds produce **peanut-shaped/ dumble cocoons**.
14. Chinese breeds produce **oval cocoons.**
15. European breeds produce **long oval cocoons.**
16. Tropical breeds produce **long pointed cocoons.**
17. Separation of cocoons based on size is called **riddling.**
18. Double cocoon reeled silk is called **dupion silk.**
19. The shape of the CSR_4 hybrid cocoon is **dumble.**
20. The shape of the CSR_2 cocoon is **oval.**
21. The percentage of the weight of shell to the overall weight of cocoon is known as **shell ratio.**
22. Indian multivoltine breeds contain shell percentage of **12-15%**, MV × BV hybrids contain **18-20%** and CSR hybrids contains **23-25%.**

23. Shell ratio (%) of univoltine and bivoltine silkworm: **15-25%.**
24. Shell ratio (%) of multivoltine silkworm: **10-12%.**
25. **Cocoon colour** is the racial character and is due to the presence of colouring pigments present in the sericin layer of the bave.
26. Cocoon colour is not permanent, washes away with the sericin during the **degumming process.**
27. Degumming refers to removal of **sericin layer.**
28. The wrinkling and irregularities on the surface of the cocoon is termed **grains.**
29. Coarse granular cocoons make poor reelability and contain **thick bave**.
30. The compactness of the cocoon is more in **middle layer.**
31. The loose-knit/ spun cocoon is referred to as **thin end cocoon.**
32. The volume of cocoon is measured by the **number of cocoons per litre.**
33. Elongation of cocoon filament is **18-23%.**
34. The process of killing pupae without affecting the structure and quality of silk is known as **stifling.**
35. **Sun drying** method of stifling is practised in Jammu-Kashmir and West Bengal.
36. Steam stifling is practised in **South India.**
37. The process by which cocoons treated in boiling water for softening the filament for easy unwinding during reeling is called **cooking.**
38. The process by which a specific number of cocoons are reeled together to produce a single thread is called **silk reeling**.
39. **Silk reeling** is the unwinding of filaments from a known number of cocoons and winding them onto a reel.
40. The building in which cocoons are reeled for the production of raw silk is called **filature.**
41. The amount of raw silk that can be reeled out from the reelable cocoons is called **raw silk percentage.**
42. Total length of filament in UV and BV races varies from **1000 m to 2000 m** and in MV Varies from **300 to 800 m.**

43. **Filament** is continuous.
44. The number of kilograms of cocoons required to produce one kilogram of reeled silk is called **renditta.**
45. The renditta of multivoltine cocoons is **8 to 14.**
46. The renditta of bivoltine cocoons is **6 to 8.**
47. Raw silk is also called as **Grege.**
48. By hand/manual sorting only **15 to 20%** of defective cocoons can be separated.
49. Non-reelable cocoons are **open-end cocoons and pierced cocoons.**
50. The specific gravity of silk is **1.36 to 1.39 g/cm^3.**
51. **1 Picul = 60kg = 1Bale = 30 Books.**
52. The size of silk filament is **Denier.**
53. The commercial measure of textile yarns is expressed as **Denier.**
54. **Warp** is vertical weaving thread/ length-wise yarn.
55. **Weft** is the horizontal weaving thread/ width-wise yarn.
56. The silk reeling waste is called **jote.**
57. The cocoon riddling machine is used for **grading cocoons.**
58. The single cocoon reeling is done with **eprouvette**.
59. Raw silk is packed as **bale.**
60. Graded silk is sold at the **Silk Exchange.**
61. Moisture absorbency of silk is up to **30% of its weight.**
62. To ensure the quality of raw silk before marketing, all the raw silk produced must be tested and graded at **Silk Conditioning and Testing House (SCTH)**.
63. Circumference of the eprouvette reel is **1.125 m.**
64. Circumference of the standard reel is **1.5 m.**
65. SMOI stands for **Silk Mark Organization of India.**
66. SMOI was launched by the Central Silk Board in the year **2004**.
67. The silk filament of multivoltine races is **finer with a little lousiness.**

68. The cocoon filaments of Korean races contain a high proportion of **flavonoids.**

69. When dry silk is kept for a specified time in a standard atmosphere having 65% humidity at 25°C, the silk absorbs only 11% of its weight of the silk raw and 10%, if it is in **degummed condition.**

70. Silk has an elongation of about **20%** of its original length before breaking.

71. **Degummed bave** has greater tenacity than raw bave.

72. Raw silk has a density of **1.33 g/cm^3.**

73. Degummed silk has a density of **1.25 g/cm^3.**

74. The electrical conductivity of silk is **poor.**

75. Different trade names for silk waste in different countries,

    Knubbs - England
    Strussa - Italy
    Frisonnette - France
    Curlies - Turkey
    Kibiso and Bissu - Japan

76. Floating system of cocoon cooking is followed in **charaka, cottage basin and multi-end reeling machine.**

77. Sunken system of cocoon cooking is followed in **automatic and semi-automatic reeling machines.**

78. In multi-end reeling machine, the reeling speed is **120-150 rpm.**

79. The process of detecting the end of the filament is called **brushing/ end grouping.**

80. The speed of winding the silk yarn on to the reel is **200 to 250 rpm.**

81. Winding speed in re-reeling is **200 to 225 m/min.**

82. For a 140 g of skein, **4 twists** in the full-length and after folding **2 twists** in the reverse form is given.

83. One book weigh about **2 kg.**

84. The tenacity of mulberry silk in the range of **3.0 to 4.5 gram per Denier.**

85. The pH of reeling water must be **6.8 to 8.4.**

86. The cocoon end gathering device is called **jetteboute.**
87. The diameter of silk filament ranges from **11 to 12 μ.**
88. The intertwining of silk filaments is achieved in **croissure**.
89. **Croissure** helps in elimination of excess water from silk filaments, combine loose filaments, to obstruct limps of filaments.
90. The croissure formed during reeling ensures better **cohesion and tenacity.**
91. The croissure used in charaka reeling is **chambon type (French type)**.
92. The croissure used in cottage and filature basin is **Tavalette type (Italian type).**
93. First country to develop ARM is **Japan.**
94. The first ARM in India was introduced at **T. Narasipura.**
95. Sericin protein contains amino acids like **large amount of lysine and arginine, and other amino acids *viz.*, serine, threonine, aspartic acid, glutamic acid, and.**
96. **15,000 gallons of water** is used to obtain 1000 kg of silk.
97. The three parts of the skein are **flower, fascia and spiral.**
98. Charaka has **2-4** reeling ends.
99. Cottage basin has **4 to 10 ends.**
100. Multi-end reeling machine has **8-10 ends.**
101. Automatic reeling machine (ARM) has **200-400 ends.**
102. Bleaching of silk is done by using oxidizing agents like **H_20_2, $KMNO_4$, sodium peroxide.**
103. Grades of silk:

    I category - **18 Denier and below**

    II category - **19 to 33 Denier**

    III category - **34 Denier and above**
104. The length of silk filament weighing one gram is **9000 m.**
105. The book weight maintained during the packaging of silk skein is **2 kg.**
106. **Serigraph** is tensile strength testing machine.
107. Elongation of the cocoon filament is **18-23%.**
108. Tenacity of dry cocoon is **3.2 -3.4 g/d.**

109. The innermost layer of cocoon which do not yield any reeled silk is **pelade layer.**
110. Degumming loss in raw silk is **22%.**
111. The twisting of two or more silks threads to give suitable strength to the yarn for weaving is known as **doubling.**
112. The process of drawing the raw silk in the form of hanks on to the bobbins is known as **winding.**
113. The process of removing the natural raw silk stain/ colour to make it fit the yarn fabric is called **bleaching**.
114. **Seriplane test** is done to evaluate the evenness, cleanness and neatness defects of raw silk.
115. The raw silk reeled from cross breed cocoons in multi-end reeling machine gives **'C' grade silk.**
116. The largest cocoon market in India is located at **Ramanagara.**
117. The second largest cocoon market is located at **Sidlaghatta.**
118. **Ushnakoti** is drying chamber of cocoons.
119. The quantity of cocoons required to produce one kilogram of raw silk is called **Kakame**.
120. The price/ cost of cocoons required to produce one kilogram of raw silk is called **Kakame cost**.
121. The solubility of **PM × CSR_2** cocoon is less than **CSR_4** and more than **pure Mysore.**
122. In Karnataka, cocoon is marketed in **Regulated Marketing System.**
123. The process of tying the skeins using the coloured cotton threads to prevent the intertangling of reeled silk is called **lacing.**
124. The regulated place where silk yarn is sold and purchased is known as **Silk Exchange.**
125. Sericin-I, II and III are present in the ratio of **2:2:1.**
126. In charaka reeling device, the tarpatti has **4** number of holes for gathering of end filaments.
127. The raw silk imported from China is **2A-3A grade.**
128. The cocoon cooking and reeling is performed in the same basin in **country Charaka.**

129. The evenness of raw silk is the prime objective of reeling in **automatic reeling machine.**
130. The optimum space between spiral folds in a skein is **5 cm.**
131. To produce 1kg of cocoon, the silkworms must be fed with **20 kg** mulberry leaves.
132. The outermost layer of sericin is **Sericin-III.**
133. When the silk fibre is impregnated with water, the strength of the silk is increased by **20%.**
134. The elongation of wet cocoon filament is **28-29%.**
135. The total volume of cocoons stifled in barrel steaming is **15-20 kg**.
136. The cocoon spun by female silkworm normally weighs **less** than male.
137. Re-reeling speed of jetteboute is between **400-600 rpm.**
138. In the floating system of cocoon cooking, the temperature maintained in reeling bath is **40-45°C.**
139. In the sunken system of cocoon cooking, the temperature maintained in reeling bath is **30-35°C.**
140. Winding test is conducted to find out **continuity of the yarn.**
141. To maximize the size deviation in raw silk, the denier of cocoon filament should be **minimum 3 denier.**
142. Neatness of silk filament is controlled by the gene **'Ne'.**
143. The quantity of raw silk produced in a cottage basin per day is **800 g to 1 kg.**
144. Silk waste per cent from cocoon reeling industry is **30%.**
145. During reeling, the tension of the yarn depends upon **angle formed by the croissure and length of the croissure.**
146. According to Chinese silk industry, in multi-end reeling the recommended croissure length is **13-18 cm.**
147. In an automatic reeling machine, the recommended croissure length is **8-10 cm.**
148. **Bisu machine** is used to separate the pupae from **pelade layer** of the cocoon.
149. After conditioning, the silk skein should contain **9.5-11%** of moisture.

150. The boil-off loss ratio for bivoltine is found to be **24%.**
151. The boil-off loss ratio for multivoltine is **28-30%.**
152. Jetteboute was invented in France by American engineer named **Serrel.**
153. Jetteboute increases raw silk productivity by **33%.**
154. The quantity of raw silk produced in charaka is **2-4 kg** by 2 operators/ 2 ends.
155. The quantity of raw silk produced in multi-end reeling machine is **12 kg/ 20 basins.**
156. The humidity of the cocoon storage room should not exceed **70%** and temperature about **27%- 30%.**
157. Steamed cocoons should be reeled within **8-10 days.**
158. For weaving carpets, silk filaments of **21-22** denier are used.
159. In parachutes, silk used is of **13-14 denier**.
160. The cluster of cocoons from which the silk thread is formed is called **rose or rosette.**
161. The conical formation of the filaments from the rosette to the threader is called as **balloon.**
162. The proportion of water required during reeling process:

     Cocoon cooking part - **7%**

     Reeling - **66%**

     Re-reeling - **1%**

     Boilers - **9%**

     Others - **17%**

Reeling Water Quality Standards:

Characteristics	Standard	Permissible Range
pH at room temperature	7	6.6 - 7.6
pH at boiling	8.6	8.4 - 9.2
Acidity (ppm)	1	0 - 10
m- alkalinity (ppm)	50	15 - 80
Total hardness (ppm)	80	70 - 90
Electrical conductivity (μ/cm)	70	35 - 350

163. **Sericin-III** is water soluble.
164. **Sericin-I** is not soluble in water easily.
165. An open pan system of cooking is followed in **charaka.**
166. The three-pan system of cooking is practised in **cottage basin.**
167. Conveyor-type cocoon cooking method is used in **semi-automatic reeling machine.**
168. Reeling water required for charaka is **100-150 μ** and in multi-end **200-250 μ.**
169. To determine the fine or coarse texture of the yarn and also the magnitude of variability of the yarn **size test** is followed.
170. To determine the degree of agglutination of cocoon filaments **cohesion test** is performed.
171. **Acetic acid** is used in the silk industry to increase the lustre of silk.
172. **Nitro-silk** is used in harmonium.
173. Silk absorbs metallic salts which increase its weight by up to 30%, this phenomenon is called **loading of silk or silk loading.**
174. Silk throwing involves three major steps - **winding, doubling and twisting.**
175. S-twist during twisting of reeled silk follows **anticlockwise** direction.
176. Z-twist during twisting of reeled silk follows **clockwise** direction.
177. The skeleton of the woven fabric is **warp** (gives basic structure to the fabric).
178. The muscle of the woven fabric is **weft** (imparts feel and appearance).
179. Each thread in the weft is called a **pick.**
180. Each thread or yarn in the warp is called an **end.**
181. The primary twist or first stage twist is **Z-twist**.
182. The secondary twist or final stage twist is **S-twist**.
183. The soft twisted or untwisted yarn is used in **warp.**
184. The high twisted yarn is used in **weft.**
185. The self-designing device in a loom is **jacquard.**

186. Power loom was invented by **Edmund Cart Wright** (Great Britain, 1784).
187. The process of applying colour to the fabric in definite patterns or designs or restricted dyeing is known as **printing.**
188. The largest selling dye is **sulphur black**.
189. Natwa is commonly employed in the reeling of **tasar** cocoons.
190. The quantity of cocoons reeled in Natwa reeling machine is **80 cocoons/ 8hrs.**
191. The elongation percentage of tasar silk is **20-25%.**
192. Tenacity of tasar silk is **1.8-2.2 g/d.**
193. Elongation of eri silk is **20-22%.**
194. Tenacity of eri silk is **3-3.5 g/d.**
195. The Bhir and Das type reeling appliances are used for **Muga** reeling.
196. Among the reeling waste, **30-35%** is used for spun silk and **20-25%** for Noil silk production.
197. Pierced and cut-open cocoons are utilized for the production of hard-spun yarn known as **matka yarn.**
198. Spun yarn is classified based on **number of counts.**

    40-80 counts - finer

    20-40 counts - moderate

    5-19 counts - coarser
199. From 100 kg silk waste, **16 kg** of spun yarn and **12kg** of noil yarn are produced in the spun silk mills.
200. **Jhuri** is the kind of yarn which is spun from unopened tasar cocoons.
201. TPI - **Twist per inch**.

8

VANYA SERICULTURE

1. The American moon moth is ***Actias luna.***
2. Anaphe silk produced by - ***A. moloneyi. A. reticulata, A. ambrizia, A. venata, A. infracta*.**
3. Anaphe, the non-mulberry silkworm construct cocoons communally and weighs about **3.5 kg.**
4. Anaphe silkworm is **univoltine.**
5. Anaphe pupa undergo diapause for **200-207 days.**
6. Production of Anaphe silk is confined to **Southern and central Africa.**
7. Asian moon moth is ***Actias maenas***.
8. The crimson red dyed apparel worn by erstwhile Roman noblemen was made of **coan silk**.
9. Coan silk is produced by ***Pachypasa otus***.
10. Silk fauna which is found in Mediterranean bio-geographical region is **Coan.**
11. Fagara silk produced by ***Attacus atlas*** (Giant silk moth).
12. Gonometa Silkworms are popularly called as **Egger moth**.
13. Indian moon moth is ***Actias selene***.
14. Mexican moon moth is ***Actias truneatipennes***.
15. Mussel silk popularly known as **fish wool**.
16. Mussel silk production is largely confined to **Toranto, Italy.**
17. Mussel silks is obtained from a bivalued Mollusca, ***Pinna squamosa***.
18. NGO mainly working in non-traditional sericulture states in India is **PRADAN.**

19. Spider silk is produced by ***Nephila madagascarensis, Miranda aurantia* and *Epeira* sp. (=*Araneus* sp.).**
20. Silk produced by *Anaphe infracta* is known as **Book.**
21. Silk produced by *Anaphe moloneyi* is **Koko and Trisnian- tsamia.**
22. Spider silk has higher durability and resistance to extreme temperature and humidity therefore used in **engineering laboratories and optical devices.**

Muga

1. Muga silk is popularly known as **Golden yellow silk.**
2. Muga silkworm - ***Antheraea assamensis.***
3. Muga silk production is endemic to the state of **Assam, India.**
4. The chromosome number of the Indian golden silkmoth, Antheraea assamensis, is **n=15.**
5. Commercially exploited muga silkworm primarily feeds on ***Machilus bombycina* (som) and *Litsea polyantha* (soalu)**.
6. During Muga silkworm rearing, **Katia and Jethua crop** are recommended for commercial purpose.
7. **Katia** is the main commercial crop and the **Jethwa** is the 2nd commercial crop.
8. The rearing of muga silkworm for pre-seed crop purpose is **Aherua and Jarua**.
9. The rearing of muga silkworm for seed crop purpose is **Bhodia and Chotua.**
10. Som host plant is considered the best for rearing muga silkworm - **Naharpatia.**
11. Rearing place of muga silkworm is **Muga Chung.**
12. Muga completes **6 number of life cycles in one year.**
13. Forewing and hind wings of muga silk moths are **brown rarely pinkish tinge.**
14. The female muga silk moths lay eggs on **Kharika.**
15. Average fecundity of muga moth during summer **100.**

16. Fecundity of muga moth from May to October is **150-250.**
17. Fecundity of muga moth in winter ranges from **80-100.**
18. During summer muga silkworm complete its life cycle within **45-50 days.**
19. During summer, larval duration of muga silkworm ranges from **28 to 38 days.**
20. During winter muga silkworm complete its life cycle within **120-150 days.**
21. During winter, larval duration of muga silkworm ranges from **51 to 65 days.**
22. The characteristic prothoracic hood markings on muga larva consisting of two prominent rectangular black marks appear in **third instar** (replaced by semilunar dark brown markings in the following instars).
23. Fully grown muga larvae measures **9 × 2 cm** and weighs **15 mg.**
24. **Makundi/ Masunda** dry leaves is recommended while cocooning for producing bright golden color cocoon and good quality thread.
25. Muga cocoon is **Golden brown or glossy white.**
26. Ideal time to harvest muga cocoon after spinning during summer and winter is **6-8 and 10 days,** respectively.
27. Muga egg characteristics - **streakless, brownish, 2.8 × 2.5 mm in size and 9 mg in weight.**
28. Newly hatched muga larvae measures about **7 × 1 mm and weighs 7 mg.**
29. Muga cocoon characteristics - **single shelled, light brown, oblong, closed, reelable and slightly flossy with a weak peduncle; it measures about 5.2 × 2.4 cm size and weight 6.3 g with soft shell of 0.5 g and 9.5 % shell ratio.**
30. Pupae of muga is copper brown measures **3.2 × 1.8 cm** and weighs **5.7 g.**
31. Muga moth approximate body length of the males is **3.0 cm** and females **3.5 cm.**
32. The area of muga moth fore wing is **1662 mm^2 in males and 1857 mm^2 in females**, and the hindwing is **1181 mm^2 in males and 1351 mm^2 in the females.**

33. Wing area in muga is smaller than that of ***A. mylitta.***
34. Wing expands is about **13 cm in males** and **15 cm in females.**
35. The structure of median cross vein in wings of *A. assemensis* is **straight.**
36. The larval tubercular setae are pointed in ***A. assamensis.***
37. Tying of semi-dried twigs together in a bunch to form cocoonage is known as **jali**.
38. The day in which most of the muga silkworms mature/ ripen is known as **Bharpok.**
39. Muga silkworm collected on Bharpok is used for **seed production**.
40. Muga silkworms are transferred from one plant to another using **Chaloni** (a bamboo tool).
41. *Apanteles glomeratus* prefers to lay eggs inside the **early instar muga larva.**
42. ***Prix striaticeps*** is a promising biological control agent used against ***Eucantecona furcellata*** pest of muga silkworm.
43. Flacherie disease of muga silkworm is also called as **Mukhlauga.**
44. Grasserie disease of muga is also called as **phularog.**
45. Peak incidence of flacherie disease in muga is noticed during **summer.**
46. Pebrine disease of muga is also called as **phutuka.**
47. Pest of muga used to control leaf roller is ***Apanteles glomeratus.***
48. The activity of Stink bug is found severe during **Bhodia and Aherua** crop of Muga silkworm.
49. The incidence of ***Exorista sorbillans*** is severe during **Jarua & Chotua crops of Muga.**
50. The most common weed found in Som and Soalu plantations in closer placing under shaded area is ***Borreria* sp.**
51. The quantity of lime and bleaching powder applied at the base of muga host plant is **10 g/plant.**
52. The recommended ratio of lime + bleaching powder applied at the base of muga host plants is **97:3.**

53. Wasps (*Vespa orientales*) causes maximum damage during **early instars** of Muga.

54. Fabric prepared from muga silk is **Mekala chadar.**

55. The silk produced by muga silkworm larvae which feed on mejankori leaves (*Litsea citrate*) is **Mejonkori silk**.

Name of muga crops and their characteristics

Crops	Month	Season	Cocoon Characters	Remarks
Katia	Oct-Nov	Autumn	Best cocoon, good for reeling, filament length - 612 m	Main commercial crop
Jarua	Nov, Dec, Jan	Winter	Poorest cocoon, filament length - 265 m	Pre-seed crop
Chotua	March-April	Early spring	Used for seeds	Seed crop
Jethua	April-May	Spring	Good cocoon, next to Katia, filament length – 546 m	Second commercial crop
Aherua	June-July	Early summer	Poor cocoon, not suitable for reeling, filament length - 460 m	Pre-seed crop
Bhodia	Aug-Sept	Late summer	Poor cocoon, difficult for reeling, filament length - 448 m	Seed crop

Duration of different stages of Muga silkworm

Stages	No. of days in summer (minimum)	No. of days in winter (maximum)
Egg laying and hatching	7	15
Larval	24	70
Pupa/ cocoon	14	55
Moth and egg-laying	2	3
Total no. of days	**50**	**150**

56. The Muga spun silk is generally termed **Juthari**, whereas machine-spun fine and coarse yarns are known as **Gicha yarn.**

57. Muga cocoons are marketed by **number.**

58. Productivity - **1000 cocoons = 250 g silk = 125 g of silk waste**.
59. Silk filament length is more in Katia **(500 m average)** followed by Jethua (400 m), Bhodia (300), Aherwa (225 m) and Jarua (204 m).
60. Shell ratio is more in **Katia** (8.39) followed by Jethua (8.27) and less in Jarua (5.49).
61. **Bhir** - primitive instrument used for muga reeling.
62. The productivity of Bhir is very **low** (20-man days per kg yarn).
63. To increase productivity silk productivity single person operated reeling machines/ instruments/ devices like Chowdhury reeling machine, Trivedi reeling machine, Bharati reeling machine, CMERS reeling machine, RMRS – I, II, III reeling machine, Amber charaka are used.
64. **RMRS - III and Chowdhury type** reeling device has the highest productivity.
65. **Muga silk plus** is an effective cooking chemical used to enhance the muga silk recovery.
66. **Bani** is a muga weft reeling machine.

Tasar

1. Tasar silkworms are reared in the **tropical and temperate zones.**
2. Tropical tasar silkworm - ***Antheraea mylitta*** (India)
3. Temperate/ Oak tasar silkworm - ***A. proylei*** (India), ***A. pernyi*** (China & USSR), ***A. yammai*** (Japan)
4. Cocoons of non-mulberry silkworms are sold at the local weekly market **Chaual.**
5. Tasar cocoons are marketed by **number**.
6. Tropical tasar cocoons are sold in local numerical lots: **Ganda (4 cocoons), Pan (80), Kahan (1280 in Bihar), 1600 in Orissa, 1000 in Madhya Pradesh and Khandi (4000 in Maharashtra).**
7. Tropical tasar egg **dorsoventrally flattened and bilaterally symmetrical** (length - 3 mm, diameter - 2.5 mm, weight – 10 mg).
8. On hatching, tasar larvae is **brownish yellow with black head** (7 × 1 mm, weighs 8 mg).

9. Size and weight of tasar silkworm larva at maturity is **13 × 2.1 cm and 50 g.**
10. Tasa pupa is dark brown measures **4.5 × 2.3 cm and weighs 10.3 g.**
11. Tasar silk moths exhibit distinct sexual dimorphism the females are bigger (4.5 cm) with a distended abdomen and narrow bipectinate antenna 1.5 cm long; the males are smaller (4.0 cm) with a narrow abdomen and broad antenna.
12. Male tasar wing span is about 16 cm; forewing area is **2121 mm^2**, hind wing area is about **1584 mm^2.**
13. Female tasar wing span is about 18 cm; forewing area is about **2350 mm^2**, hind wing is about **16mm^2.**
14. Tasar pupae undergoes diapause up to **5-7 months.**
15. Oak tasar larva weighs about **20 g.**
16. The shape of silvery white lateral shinning spots on tasar silkworm body is **oval and triangular.**
17. Pitcher method of rearing young age silkworms is practiced for **tasar.**
18. The voltinism of majority of the eco races of the tropical tasar is **bivoltine.**
19. Female oak tasar silk moth has triangular shaped **crepts** on wings.
20. The length of tasar cocoon peduncle is more in **Daba.**
21. Another name of *Antheria mylitta* is ***A. paphia***.
22. Japanese tasar silkworm *Anthereaea yamami* produce **green silk.**
23. Commercially exploited tasar eco race is **Daba.**
24. *Antheraea proyeli* is the interspecific cross between ***A. pernyi* and *A. roylei.***
25. Wild cocoons of eco race Railey and Modal weighs about **25-30 g.**
26. *Hardwickia binata* is a secondary food plant of **tropical tasar.**
27. Size of the tropical tasar egg is **more** than the oak tasar egg.
28. The yarn spun out from tasar cocoon peduncle is **balkal.**
29. Non-diapause and diapause seed cocoons of tasar can be preserved for about **5-7 months and 25-30 days**, respectively.

30. **Hammock formation** is first step in tasar cocoon construction process.
31. The time required for hammock formation during tasar cocoon spinning is **6 hours.**
32. In tasar silkworm, formation of ring and peduncle is completed **within 10 hours.**
33. The male oak tasar silk moth has **oval shaped eyespots on wings.**
34. The seed cocoons of tasar are preserved in the form of **garlands.**
35. Placing of Nerium flowers (*Thevetua peruviana*) inside the tasar egg incubation tray is known to increase hatching of eggs by about **8-10%.**
36. To rear 225 DFL's of tasar silkworm up to 2nd instar require **70 number** of *Terminalia* bushes.
37. **10 DFL's/ Pitcher** to be brushed in pitcher rearing.
38. Developing tasar silkworm larvae are mounted along with twigs of the host plant at **junction of branches where canopies intersect.**
39. Under tropical conditions, the orientation of tasar grainage is always **North-South.**
40. During summer tasar seed cocoons are preserved in grainage at **< 35 °C and 40-60% RH.**
41. Tasar male silk moth starts emerg **1-2 days earlier than females.**
42. Tasar moths emerge between **6-10 PM.**
43. During coupling of tasar moths the ratio of male and female moths maintained at grainage is **3:1**.
44. Prick method of mother moth examination is followed for pebrine detection in **tasar silkworm.**
45. Pricking will be done on **5th and 6th abdominal segment** of moth.
46. Daba eco race of tasar is normally **bivoltine.**
47. Sukinda eco race of tasar is **trivoltine.**
48. Two parallel brownish lines are found on the egg surface of **tasar.**
49. **Jeevan Sudha** is generally sprayed on tasar silkworm to control viral disease.

50. Recommended dose of Jeevan Sudha to rear 100 DFL's of tasar silkworm is **150 g/ 20 litre water.**
51. The disinfectant recommended for tasar egg sterilization is **Depuratex.**
52. In tasar silkworm the tubercles are distributed on **dorsal, upper lateral & caudal region.**
53. Structure of median cross vein in wing is wavy in ***Antherea mylitta.***
54. Tropical tasar larva rest underneath leaves during **hot sun, moulting and when they start spinning.**
55. Tropical tasar larva feeds from **margin to center of the leaf.**
56. Young age tropical tasar silkworm larva take rest on **leaf margin.**
57. The structure of median cross vein on wings is convex in ***A. pernyi* and *A. roylei.***
58. Chromosome number of *A. roylei* is **n = 31.**
59. Chromosome number of *A. yamamai* is **n = 31.**
60. Chromosome number of *A. pernyi* is **n = 49.**
61. Fungal disease in tropical tasar larvae is noticed particularly during **Sept-Nov.**
62. Causal agent of fungal disease is ***Pencillium citrinum* and *Pacilomyces variatii.***
63. Peak incidence of *Sycanus collaris* on tropical tasar silkworm is observed during **August.**
64. Tropical tasar moth lays eggs in batches of **5-10.**
65. Bacterial disease in tropical tasar silkworm caused by **Gram '+ve' *Bacillus* sp. and Gram '-ve' *Micrococcus* sp.**
66. Tropical tasar silkworm consumes on an average **300 g of leaf** in its life cycle.
67. During breeding, the paired Oak tasar moths are placed in **Monia.**
68. Temperate tasar silkworm is **Univoltine and Bivoltine**.
69. Fibroin of tasar silk **more resistant** to acids and alkali compared to mulberry silk.

70. Number of tubercles in tasar larva remain constant until **3rd Instar.**

71. Tubercle number **decreases** as number of abdominal segments increase in tasar larva.

72. In *A. yamami* both egg and pupal stages undergoes **diapause.**

73. Pebrine disease in tropical tasar silkworm is caused by ***Nosema mylittanis.***

74. Standard wild silk thread is made up from **eight cocoons.**

75. Total number of tubercular setae in

A. pernyi, A. roylei, A. assamensis and *A. polyphenes* - **462**

A. mylitta and *A. sivalica* - **502**

A. frithii – **506**

76. Tubercular setae is pointed in ***A.mylitta* and *A. polyphenes.***

77. Tubercular setae are knobbed and pointed in ***A. roylei* and *A. pernyi.***

78. Wing venation patters in *Antheraea* sp.:

Silk moths	Wing Venation Pattern
A. pernyi, A. roylei	Convex
A. polyphenus, A. yamamai	Concave
A. mylitta, A. sivalica, A. frithi	Wavy
A. assamensis	Straight

80. The reelability of tasar cocoon is **50%.**

81. Average filament length of tasar cocoon is **800 m.**

82. Hand prepared yarn out of tasar cocoons is called **Gicha.**

83. Tiger band disease is noticed in **oak tasar silkworm.**

84. Tiger band disease caused by **Apha baculovirus (*A. proylei* Nucleopolyhedrosis virus).**

85. Silver shining spots on tasar silkworms appear during the 3rd instar from **2nd to 7th abdominal segments.**

86. Tasar larva possess **5 types** of tubercules (dorsal, upper lateral, lateral, lower lateral, caudal).

87. The larval duration of tasar silkworm is **26-28 days in 1st crop; 42-45 days in 2nd crop; 55-60 days in 3rd crop.**

88. Prothoracic hood of the 1st instar, tasar larvae dorsally bears on oval black spot, which early in the 2nd instar becomes 'M' shaped later on 'V' shaped with 2 dots, these are absent in 3rd instar, but appear in the 4th and 5th instar as two semi lunar red markings.
89. Scientific name of Asan is ***Terminalia tomentosa***.
90. Scientific name of Arjun is ***Terminalia arjuna***.
91. Scientific name of Sal is ***Shorea robusta.***
92. **Tasar Keet Ousadh (TKO)** is effective against all diseases.
93. *A. proylei* eggs are **brownish** in color.
94. **Sukinda** eco race shining spot are completely absent.

Silkworm		Wing span	Fore wing	Hind wing
Eri	Male	13cm	1242 mm^2	890 mm^2
	Female	15cm	1465 mm^2	1037 mm^2
Muga	Male	13cm	1662 mm^2	1181mm^2
	Female	15cm	1857 mm^2	1351 mm^2
Tasar	Male	16cm	2121 mm^2	1584 mm^2
	Female	18cm	2350 mm^2	1850 mm^2

Eri

1. Eri is derived from Sanskrit word **'Eranda'** and Assamese word **'Era'**, which means castor plant.
2. Eri silkworms, *Samia cynthia ricini* are **multivoltine.**
3. The chromosome number of *Samia cynthia ricini* is **2n = 27♀/28♂**.
4. *Samia cynthia ricini* is derived from its wild form ***Samia canningi.***
5. The primary host plant of eri silkworm is **castor (*Ricinus communis*).**
6. Secondary food plants of eri - ***Heteropanax fragrans* (Kesseru), *Manihot utilissima*, *Carica papaya*, *Evodia fraximyalia*, *Ailanthus excelsa*, *Plumia rubra***.
7. Castor belongs to family **Euphorbiaceae.**
8. Inflorescence type in castor is **terminal raceme.**
9. Brick red colour cocoons are produced by **Kokrajhar** eco race.

10. The late-age eri silkworm rearing is carried out by **bunch rearing, tray rearing, platform rearing.**
11. Eri silkworm tubercles are distributed on **dorsal and upper lateral region.**
12. Eri silk moth has **bipectinate antennas.**
13. Eri silkworm completes its life cycle within **44 days** in summer and **85 days** in winter.
14. A conspicuous black spot pterostigma with white tinge at the wing apex is feature of **eri silk moth.**
15. Best season for commercial eri rearing is **September-November.**
16. Prothoracic hood of eri silkworm of the 1st instar larva has a black dorsal band which splits up into a pair of crescent shaped marking in the 2nd and 3rd instar; the marking disappears at the 4th instar.
17. Eri silkworm spins **open end** cocoons (hence called ahimsa cocoons).
18. Eri eggs are ovoid, candy white measure **1.5 × 1.0 mm and weighs 6 mg.**
19. Early instars of eri silkworms are greenish yellow measures **5 × 1 mm and weighs 1.5 mg**.
20. Third instar onwards, the eri silkworm body colour segregates into **yellow, cream, green, blue or white.**
21. The fully mature eri larvae measures **7.0 × 1.5 cm, weighs 8 g and is translucent covered with a white powdery substance**.
22. A whitish powdery substance is present on eri silkworm body is **urate crystals.**
23. Eri pupae measures about **2.8 × 1.5 cm and weighs 2.6 g.**
24. Cocoons are open and unreelable measures **4.0 × 2.5 cm, weigh 3 g, exhibit colour polymorphism, brick red or creamy white, shell weight and shell ratio are 0.4 g and 13%, respectively.**
25. Eri moth measures **2.5 cm (male) and 3.0 cm (female)** in length.
26. Wing expanse of male and female eri moth is **13 cm and 15 cm,** respectively.

27. Eri moth forewing area is about **1242 mm^2 in males and 1465 mm^2 in females**; hind wing area measures **890 mm^2 in males and 1037 mm^2 in females.**
28. The ocellus in both the sexes of eri moth is **crescent** shaped.
29. **Moth emergence starts from 3:00-5:00 in the morning and may continue until 16:00- 17:00 hours with maximum emergence in the hours 3:00 to 9:00.**
30. Coupling duration in eri grainage is **3 to 6 hours.**
31. Fecundity of eri moths **350-400 eggs.**
32. Eri silk is characterized by **low sericin** (4.96%) (mulberry silk - 10%, tasar silk - 8.62% and muga - 7.88%).
33. Eri fibroin has a **higher** alanine than glycine content by the ratio of 0.6:1.
34. Eri silk is used for preparing warm clothes like **Chadars.**
35. British called eri silk as '**Palma Christi Silk**', was woven to heavy clothes called **Bar Kapoor.**
36. Eri silkworms can be raised on plains as well as at high altitudes upto **1500 MSL.**
37. The temperature ranging between 15°C in winter and 35°C in summer and RH ranging from 50-100% are conducive to ericulture, and optimum range is 24-28°C and 70-80%, temperature and RH, respectively.
38. Eri silk is also known as **poor man's silk or Ahimsa silk.**
39. Male eri worms ripe **earlier** to the female's ones.
40. **Kharika** is used for egg laying.
41. Completely domesticated non-mulberry silkworm is **eri.**
42. Bunch feeding of leaves is very common in **eri silkworm rearing.**
43. White colour eri cocoons are produced by **Borduar and Titabar eco races.**

9

MODEL PAPERS

Model Paper: 01

1. The process of rearing a silkworm for silk is known as **Sericulture.**
2. The process of growing mulberry plants for silkworm cocoon production: **Moriculture.**
3. A Mulberry silkworm feeds on ***Morus*** species.
4. There are four types of silk in India, namely: **Mulberry silk, Tasar silk, Eri silk and Muga silk.**
5. Status of Karnataka Silk production in India: **First position.**
6. Commercially grown Silks of Andhra Pradesh: **Mulberry and Tasar**
7. Status of Andhra Pradesh Silk production in India: **Second position.**
8. APSSRDI a pioneer Sericulture Institute of Andhra Pradesh is in: **Hindupur.**
9. Rayalseema contributes for about % of Silk production of Andhra Pradesh: **90%.**
10. Central Silk Institute for Muga and Eri research is located at: **Ladoigharh, Assam**.
11. Silk gained entry to India through: **Tibet.**
12. Major consumer of silk in the world is: **India**
13. Mulberry variety recommended for shade condition is **Sahana.**
14. Biofertilizer that Fixes nitrogen in Mulberry ***Azatobacter.***
15. Mulberry inflorescence is called **Catkin.**
16. Most suitable soil type for mulberry is **Red loamy Soil**.
17. Mulberry plant is **Highly cross** pollinated.
18. Monophagous nature of mulberry leaf is due to the biting factor **Morin**.

19. Basically, mulberry is **Tree**.
20. Perennial natured Mulberry plant has **Taproot system**
21. Farm yard manure recommended per hectare of mulberry per year is **20 tonnes**
22. Papaya Mealy Bug menace in mulberry can be managed by release of predator ***Acerophagus papaya***
23. Release of Bio-agent recommended for management of leaf roller in mulberry ***Trichogramma chilonis***
24. Mulberry variety recommended for mildew resistant is **MR 2**
25. Insect responsible for causing Tukra disease is **Mealy bug**
26. Typical symptom of Bihar Hairy Caterpiller (*Spilosoma obliqua*) infestation is **Leaf skeletonization**
27. A broad-spectrum Predator recommended for effective management of Thrips ***Chrysoperla*** spp.
28. Plant based formulation for Root knot management is **Nemahari**
29. ***Diaphania pulverulentalis***- leaf roller damage in mulberry is usually seen in Top leaves
30. Root rot disease in mulberry is caused by **Fungi**
31. Recently recommended multi nutrient foliar spray for improving mulberry leaf quality by CSRTI, Mysore is **Poshan**
32. Management of Tukra can be effectively done through release of Both ***Cryptolaemus montrouzieri & Scymnus coccivore***
33. Multivoltine hybrid developed by **APSSRDI is APM1 X APS8 (Swarnandra).**
34. Number of ovarioles in *Bombyx mori* adult are **8**.
35. Number of pairs of abdominal legs in *Bombyx mori* are **5**.
36. Silkworm have spiracles as respiratory openings are **9.**
37. Mouth parts in silkworms are **Hypognathus.**
38. Dorsal caudal horn in silkworm body is found on Dorso median line of **8th abdominal segment.**

39. Silk gland is modified from **Labial gland.**
40. Serious pest of mulberry which is commercially exploited is **Bombyx mori.**
41. Silkworm *Bombyx mori* has Secondary hair as **setae.**
42. Pupa of Bombyx mori is **Obtect.**
43. In producing double hybrids, FC refers to **Foundation Cross.**
44. CSR-2 is a popular **Bivoltine breed.**
45. In silkworm rearing micro climate is monitored using **Thermo hygrometer.**
46. Botanical based silkworm bed disinfectant is **Ankush.**
47. Optimum temperature & Relative Humidity required for chawki/young silkworms are **26-28 °C & 85-90%.**
48. Double cocoon formation is largely due to **High density of mounting**.
49. Double cross hybrids have **Four parents.**
50. Orientation of Silkworm rearing house should be **East-West.**

Model Paper: 02

1. Marketing of Bivoltine cocoons should be done on **8th Day.**
2. Number of worms mounted on chandrike / mountage per square feet **40-50.**
3. Egg attains pin head stage before hatching at **48hrs.**
4. In Silkworm rearing Hatching is a **Photo-periodic response.**
5. Silkworms prefers more for its activity in **Dim light.**
6. Rearing space required for Bivoltine Silkworm rearing is **800-900 sq.ft**
7. Hormonal formulation used for uniform maturity of silkworms is **Sampoorna.**
8. Bivoltine hybrid recommended for irrigated areas of Andhra Pradesh **$CSR_2 \times CSR_4$.**
9. In Silkworm seed production, Basic seed means **Parental seed.**
10. Head quarter of National Silkworm Seed Organization is at **Bangalore.**

11. In Silkworm egg production centres natural moth emergence is allowed for **Multivoltine Cocoons.**
12. Preservation of male moths is done for **Second pairing.**
13. Black Boxing of Silkworm eggs is done to achieve **Uniform hatching**.
14. Pairing duration for Silk moths is **3-4 hours.**
15. Quantity of loose eggs packed and its weight is **50 DFLs & 18 grams.**
16. Mother Moth Examination is done to detect **Pebrine.**
17. Silkworm Seed Technological Laboratory (SSTL) under CSB is located in **Bangalore.**
18. CSR Hybrids are developed under the technical guidence of **JICA, Japan.**
19. Grasserie disease in silkworms is caused by **Borrelina.**
20. Calcified cocoons are the symptoms of **Fungus**.
21. Eco-friendly botanical based formulation to supress Grasserie & Flacherie is **Amruth.**
22. Flaccid condition in silkworms is due to combined infection of **Bacteria & virus.**
23. Green muscardine disease is caused by ***Metarhizium anisoplea.***
24. For effective and healthy silkworm rearing, disinfection per crop is recommended **Two times.**
25. Multivoltine race resistant to diseases, used as female parent in Kolar gold Cross breed is **Pure Mysore.**
26. In 1960's Sericulture vanished in France due to **Pebrine.**
27. Uzi fly a major pest of silkworm, here Uzi is the name of place, where it was first noticed is in **Japan.**
28. In south India Uzi fly menace was first noticed in **Karnataka**.
29. Uzi fly belongs to the family **Tachinidae.**
30. Uzi fly is a **Regular pest.**
31. Uji maggots undergoes **3 instars.**
32. *Nesolynx thymus* parasitises **Uzi pupa**.
33. Uzi powder acts as a **Ovicide.**

34. The grainage pest on stored cocoons is **Dermastid beetle.**
35. Unit to measure the size of the cocoon is **Number/litre**.
36. Single cocoon weight of bivoltine hybrids is **1.8-2.0 g.**
37. Silkworm breeds of Chinese origin spin **oval cocoons**
38. The cocoon shell ratio of multi x bivoltine hybrids ranges from **18-20%.**
39. The purpose of cocoon stifling is to **Kill pupae**.
40. The quantity of cocoons required to produce one kilo of raw silk is called **Renditta.**
41. The extent of Fibroin in cocoon is **75-80%.**
42. The shape of CSR 4 cocoon is **Dumbel.**
43. Cooking of cocoon is done during **silk reeling for Softening of sericin and easy unwinding**.
44. Cocoon riddling machine is used for **Grading of cocoons by size.**
45. The floss content is more in the cocoons of **Pure Mysore.**
46. ARM stands for **Automatic Reeling Machine.**
47. Silk filament is technically known as **Bave.**
48. Croissure formed during reeling ensures Both **Better Cohesion and Better tenacity**.
49. The reel speed of multi-end reeling machine for Bivoltine cocoons is **120 m/min.**
50. The standard pH of the boiling water during reeling is **8.6.**

Model Paper: 03

1. Phosphorus is Gummy component of cocoon shell is **Sericin.**
2. Reelable defective cocoons are reeled on **Charaka.**
3. Dupion silk is obtained from **Double cocoons.**
4. Chambon type of croissure is present in **Charaka.**
5. Silk waste percent from cocoon reeling industry is **30 %.**
6. Perimeter of Reels in Multi-end reeling machine is **75 cms.**

7. SCTH stands for **Silk Conditioning & Testing House**.
8. Pierced cocoons are used to get **Spun silk.**
9. The raw silk in the form of skein is packed as **Bale.**
10. Which of the feature is the most important in quality silk reeling Raw Silk: **Denier**.
11. Central Silk Technological Research Institute is located at **Bangalore.**
12. The small skein approximately should weigh **70g.**
13. Evenness test is done using **Seriplane.**
14. In Andhrapradesh SCTH (Silk conditioning and testing house) is at **Dharmavaram**.
15. Auto sorter is the machine used to measure **Size of silk.**
16. Vanya silk grown in Andhra pradesh is **Tropical Tasar.**
17. Costliest silk is **Muga silk.**
18. Som & Soalu are the host plants of **Muga silkworm**.
19. Production of cocoonase enzyme is absent in **Eri silk moth**.
20. State producing both Eri and Muga in larger quantity **Assam.**
21. The Indian temperate tasar is the cross between ***Antheraea pernyi* X *Antheraea roylei***
22. The multivoltine non-mulberry silkworm is **Eri.**
23. Primary host plant of tropical tasar is ***Terminalia spp.***
24. Central Tasar Research & Training Institue of CSB is located at **Ranchi, Jharkhand.**
25. In A.P, Basic Seed Multiplication & Training Centre for Tasar is located at **Rampachodavaram.**
26. Major vanya silk produced in India is **Eri.**
27. Mulberry tea is preferred for regulating **Diabetes.**
28. Mulberry fruit is rich in **Vitamin C.**
29. In India **Eri pupa** is relished as protein rich food in Assam
30. Internal Sutures are made using **High quality braided silk.**

31. Sericin is extracted by **High Temperature & high Pressure.**
32. Fibroin is used for **Health products and wound healing.**
33. Katia a kind of silk extracted out of **Ring & Peduncle of tasar**.
34. Gicha silk is produced using **Mud pot**.
35. Spun silk mills are concentrated in **West Bengal & Assam.**
36. **Noil yarn** is by product of spun silk.
37. Majority of the Silk Producing Insects belong to family Bombycidae of order **Lepidoptera.**
38. The silkworm commonly found in Indian Tropical Tasar is ***Antheraea mylitta***
39. The Indian foreign exchange generated through export of Silk in 2021 was **211 million dollars.**
40. Raw silk production in India during 2020-21 was **33770 MT.**
41. Which state is the Leading producer of Silk in India: **Karnataka.**
42. The first authentic reference of silk production is found in **China.**
43. The Government of India through Central Silk Board (CSB) has implemented a Central Sector Scheme named as **Silk Samagra.**
44. According to some Historians, Silk was exported from India to Rome during Period in **58 B.C by Kanishka.**
45. Family of mulberry is **Moraceae.**
46. Mulberry plants are the native of **Lower range Himalayas either of India or China.**
47. White mulberry (or Eastern Asia Mulberry) ***Morus alba.***
48. Which of the following mulberry varieties are suggested for heavy rainfall regions: **S-795.**
49. Fertilizers should be applied at least at days after planting of the mulberry is **55-60 days.**
50. Cuttings are prepared in India from young plants are called **soft wood cuttings.**

Model Paper: 04

1. The following are the cultivation practices of mulberry tree asexual propagation are **seedling, grafting and budding.**
2. Root hormones/growth regulators are used to induce rooting in the "cutting method of propagation is **Indole Butyric Acid.**
3. Grafting is a method where **Scion** is attached to the stock.
4. For grafting of mulberry, the scion used should be a well grown shoot of **one** year old plant.
5. Method is used to maintain the characteristics of the selected variety: **Grafting.**
6. The method of propagation which involve the development of roots from a stem while still attached to the mother plant is called **Layering.**
7. In **air Layering** method 1-2 cm circular bark is removed from middle of the branch and used for propagation of mulberry
8. Planting system of mulberry is convenient for mechanical operations **Single row** in low density.
9. Spacing system allows high productivity of mulberry **Pit system**
10. Planting system followed in South India for irrigated mulberry crop is **Pit system**
11. For mulberry cultivation, which of the following elements constitute major nutrients and required in larger amount **C, H, O, N, P, K**
12. Periodically and methodically cutting of some branches of mulberry is called **Pruning** and **Training**
13. For rainfed mulberry crop, pruning is given **Once** in a year.
14. The harvested leaves of mulberry Contains high quantity of **Starch**.
15. During **leaf picking** method of harvesting, the terminal buds are nipped off for rapid development of lateral shoots
16. Whole shoot harvest is fed directly after **Third moult** of silkworm
17. Fungal leaf spot of mulberry is caused by ***Cercospora* spp**
18. Which of the following is the vector of Tukra disease of mulberry **Mealy bug**

19. Powdery mildew of mulberry is caused by ***Phyllactinia corylea***
20. For prevention of root disease, the root saplings can be dipped in **0.2% bavastin solution**
21. Biocontrol agent can be used against root rot fungus in mulberry is ***Pseudomonas***.
22. Sudden wilting, withering and defoliation of leaves of mulberry is the symptom of **Root rot**
23. Disease is severe in high moist soil is **Collar rot**.
24. The fungicide is used to manage stem canker of Mulberry is **Carbendazim**
25. The greenish black lesions that appear in initial stage on stem of mulberry and later changes to dark black hard carbonaceous structures are the symptoms of **Stem canker**
26. Most common species of nematode affecting the mulberry is ***Meloidogyne incognita***
27. The preventive measure for nematode management is **Deep summer ploughing**
28. The immature stages of nematodes secrete proteolytic enzyme which releases growth hormone **IAA** and causes **Hypertrophy** in the roots.
29. Lepidopteran pest attacking mulberry is ***Spilosoma obliqua***
30. Which of the following pest of mulberry is not recorded in India is **Mulberry tiger moth**
31. Termites infesting in mulberry is ***Odontotermes obesus***
32. Which of the following is/ are cultural methods of management- **Rouging, Sanitation, Proper Pruning**
33. Clipping of pest affected apical part is the control measure for **Leaf roller**.
34. Use of *Trichoderma* spp is **biological method** of management
35. Use of light trap to attract the adults of *Spilosoma obliqua* is **Physical method** of management.
36. Silkworm belongs to order **Lepidoptera**
37. Number of moults in *Bombyx mori* is / are **four**
38. The antennae of *Bombyx mori* for males and females are **Pectinate** for both sexes

39. Which of the following races of Silkworm lay only diapausing eggs **Univoltine.**
40. Which of the following races of Silkworm lay only non-diapausing eggs **Multivoltine.**
41. The eggs of silkworm are **Centrolecithal**
42. The mature egg of *Bombyx mori* is surrounded by **Chorion** and **Vitelline membrane**.
43. The collection of newly hatched larvae of silkworm from the egg cards is called **Brushing**.
44. The silk glands of silkworm are **Exocrine**
45. Which of the following is a useful bed disinfectant in sericulture, especially against viruses is **Slaked lime**
46. General preventive measures for hereditary pebrine disease in silkworm include surface sterilization of disease-free egg by dipping egg cards in **2% Formalin** solution for 20 minutes and then washing in running water.
47. Which of the following disinfectant is a strong oxidizing agent and used for all type of rearing houses of silkworm **Chlorine dioxide**
48. The rearing of 1st and 2nd instars of silkworm falls under **Chawki rearing**
49. Brushing should be done with the help of **Soft feather**.

Model Paper: 05

1. Brushing should be done with the help of **Dried mulberry** leaf powder
2. Which of the following chemical should be placed in the rearing rooms of silkworm during rainy to absorb excess moisture **Unhydrated Lime**
3. **Late age silkworm** rearing include rearing of worms from third to last instar
4. Maximum food (approximately 94%) is consumed by silkworm during **Late age silkworm rearing**
5. The optimum spacing (density) for Chandrike, popular Indian mountage is **50 worms per 0.1 m^2**
6. The requirement of **leaf quality** is different for young and late age larvae of silkworm

7. The requirement of **temperature** is different for young and late age larvae of silkworm
8. The requirement of **humidity** is different for young and late age larvae of silkworm
9. During moulting, care should be taken and dusting should be done especially during **Rainy season**
10. Three different types of factors are involved in the feeding behaviour of *Bombyx mori*, they are **Attracting, Biting and Swallowing**
11. During moulting periods, the rearing bed has to be dried by dusting **Lime powder**
12. **Mountage** is used to enable fully mature larvae of silkworm to spin cocoon.
13. Normally, the cocoons of Bombyx mori should be harvested on **5th or 6th day** of spinning.
14. Central Silk Board is situated at **Bangalore**.
15. Bivoltine races has **Two generations per year**.
16. Muga and Eri silkworm are **Multivoltine**.
17. The grainage in sericulture should be done **Away from industrial area.**
18. Eggs of silkworm are subjected to surface sterilization by 2% Formalin. **To avoid any disease spread before packing.**
19. **Cocoon cutting** is carried out for sex separation of pupae for hybrid seed production of silkworm.
20. Egg diapause of silkworm can be broken artificially by **Acid treatment**.
21. Artificial hatching of silkworm eggs in cold storage with the common acid treatment takes **20 days**.
22. The strain of Nuclear Polyhedrosis Virus that infect mulberry silkworm is **NPV Bm**.
23. Silkworm become sluggish, loose appetite, become immobile and discharge soft excreta. These are the symptoms of **Grasserie disease.**
24. Causative organism of septicaemia in *Bombyx mori* is ***Bacillus, Streptococcus and Staphylococcus***

25. Which of the following disease is caused by Toxicosis of bacterium in silkworms **Sotto disease**
26. A farmer is practicing sericulture, who observed that some silkworm larvae are showing sluggish movement, decreased appetite, straightened body, swollen thorax, shrinkage of abdominal segments, vomiting, loss of clasping power of legs, soft and discoloured body and soft or bead like excreta: **Bacterial Septicaemia.**
27. ***Beauveria bassiana*** causes disease in silkworms Muscardine
28. The infection of *Beauveria bassiana* starts through **Integument**
29. Symptoms of infected larvae of silkworm are loss of appetite and sluggishness, Larvae die within 3-5 days of infection. Dead larvae initially appear flabby but harden later due to complete intrusion of the pathogen. The disease is **Muscardine.**
30. Pebrine is a **Protozoan disease**
31. Which of the following disease in silkworm was the major cause of decline in silk industry in Europe especially in France in the latter half of the nineteenth century? **Pebrine**
32. Prevention of hereditary pebrine disease is to use disease-free seed prepared by **Pasteur Method**
33. Uzi fly belongs to **Diptera** order and **Tachinidae** Family
34. **Uzi fly** is a parasite of larval stage of *Bombyx mori*
35. **Uzi fly** is a parasite of pupal stage *Bombyx mori*
36. Uzicide which is used to control the population of Uzi fly in sericulture, is a **ovicidal formulation**
37. Dermestid beetles feed upon **Cocoons** of Silkworm
38. The antennae of Dermestid beetles is **Clubbed**
39. Which of the following chemical is in maximum amount in cocoon of silkworm **Fibroin**
40. Number of cocoons required (in kg) to obtain 1 kg of reeled silk is called **Renditta**
41. Unit used to denote the thickness of silk is called **Denier**

42. Multivoltine races of silkworm have renditta value of **8-14**
43. **Cocoon cooking** is the process to allow Penetration of water in the cocoon
44. Which of the following reeling predominate in Indian Sericulture industry **Charakha reeling.**
45. The reels of multi-end machine are of small size with circumference of **60-75 cm**
46. **Jetteboute** is a cylinder made up of two concentric tubes Metallic
47. The agglutinating and twisting of the reeled thread is done by **Croissure**
48. Two type of Croissures used in India are **Chambon type and tavellette type**
49. **Jetteboute** is used as thread guide in reeling machine
50. **Jetteboute** is used for better attachment of threads

Model Paper: 06

1. **Jetteboute** is used for the removal of slugs and waste
2. Stop motion device is not present in **Simple reeling machine**
3. Stop motion device is present in **multi-end reeling machine, Automatic reeling machine and Multi-end reeling machine.**
4. The defects of the reeled silk can be eliminated through **Re-reeling**
5. Reeling velocity is measured as the length of raw silk reeled in **1 minute**
6. Twisting of raw silk is called **Skeining**
7. **"Winding"** is a Mechanical test for raw silk
8. **Cohesion test** is used for Raw silk testing
9. A dupion cohesion tester consist of a framework designed for continuous thread of raw silk placed in a **Zig - zag manner.**
10. **"Large slug defects"** in the reeling of silk can be avoided by Low temperature drying
11. Standard Grading is based on grading formulated by **Bureau of Indian Standards based on International Silk Association Standards.**
12. **Spun silk fabrics** are made up of silk waste generated during reeling of the cocoon

13. Muga silk is in colour **Golden yellow**
14. The domesticated Eri silkworm is **Multivoltine**
15. Which of the following is wild Eri silkworm ***Bombyx Cynthia***
16. Muga silkworm feeds on Leaves of **Som and Soalu trees**
17. Indian temperate Tasar silkworm (or Indian Oak Tasar) is the hybrid of **Chinese *Antheraea pernyi* and Indian *Antheraea roylei***
18. Muga silkworm is endemic to India and occurs in **Brahmaputra Valley** and **adjoining hills of Assam.**
19. Tasar cocoons are cooked with **Hydrogen peroxide and soda ash**
20. Eri cocoons are **Open mouthed and with discontinuous silk filament**
21. The leaf maturity of mulberry is measured by estimating the **Calcium ratio**
22. In terms of medicinal value, the Mulberry leaves are considered to be **Diaphoretic**
23. Fresh mulberry fruits Contains **Calcium**
24. **Enzymatic hydrolysis** is commonly used to increase juice yield of mulberry fruits
25. **Waste products** of silk reeling could be utilized to produce value added products
26. The **anthocyanin content** of mulberry wine is higher than red grape wine
27. The first species of Lepidoptera to have a completely sequenced genome is ***Bombyx mori***
28. Book "The Silkworm an Important Laboratory Tool" is written by **Yataro Tazima**
29. Juvenile hormone in silkworm is secreted by **Corpus allatum**
30. The short filaments deposited during dressing process which cannot be spun are called Noil **yarn.**
31. The fat content of silkworm pupae is on dry basis is **30.1%.**
32. Tasar waste material contains significant amount of Gummy material which hinders mechanical processing in **spun silk mill.**

33. Double cocoons are particularly reeled to produce silk which has high commercial value is **Dupion.**
34. Moths emerged from double cocoons may be used for preparation of commercial **disease free layings (DFLs) of multi X bivoltine.**
35. Pupal oil of the silkworm pupae can be obtained by **Hexane oil extraction method**
36. Silkworm Fibroin can be solubilized by **Degumming**
37. Raw silk thread obtained from cocoons contains **20-30%** sericin protein
38. Medicinal and cosmetic application of sericin include **Anti-aging and anti-wrinkling**
39. The silkworm larva ceases to eat and begins spinning silk all over its body from **outside to inside**
40. This species produces silk of the superior quality ***Bombyx mori***
41. Silk is produced by **Late age larva**
42. **Honey, silk and lac** are secretory substance of insects
43. What is the life span of an adult *Bombyx mori* **3-4 days**
44. The egg of silk moth is **Centrolecithal.**
45. *Bombyx mori* belongs to the class **Insecta.**
46. The life cycle of mulberry silkworm is completed in **30 days**
47. Eggs of silk worm are kept in cold storage at **5^0C**
48. The silk fibre is held together by substance known as **Sericin**
49. Mullbery, muga, oak, and tasar all the four varieties of silk arc present in single country, the country is **India**
50. Sericulture is native of **China.**

Model Paper: 07

1. The process of shedding down of exoskeleton in silkworm is called **Moulting.**
2. Viral disease of silkworm is **Grasserie**
3. Queen of textiles **Silk**

4. Muscardine disease have causative agent is ***Beauveria bassiana***
5. Silk contains a protein known as **fibroin**
6. How many stages are there in the life cycle of a silkworm - **04**
7. Which of the following is considered as a silk city of India **Jaipur, Kanchipuram, Pochampally.**
8. Eri silk: **Also known as Arundi silk, produced in Bengal, Bihar and Assam, Cocoons are soft and white.**
9. How many stages are there in the life cycle of a silkworm- **Four**
10. Cocoon is formed by how many pairs of glands- **Two**
11. **Pebrine disease** in silkworms is not caused by a fungus.
12. Paheli wanted to buy a gift made of animal fibre obtained without killing the animal. Which of the following would be the right gift for her to buy **Woollen shawl**
13. Silk fibre is obtained from **Cocoon.**
14. Wool fibre can be obtained from which of the following: **Goat, Llama, Alpaca**.
15. **Selective breeding** is a process of Selecting the parents with desired properties.
16. **Reeling of silk** is the process of taking silk threads from cocoon.
17. **IAA** is a growth regulator
18. **Kolar** is the highest silk producing district in Karnataka
19. All the silkworms have **hypognathous head**.
20. **Silk** is a natural fibre
21. **Diapausing hormone** is secreted by sub-oesophageal ganglion.
22. **Sericigenous larvae** have fixed number of Malpighian tubules.
23. A sudden heritable change is called **mutation.**
24. Frontal nerves connect to **hypocerebral ganglion**
25. M-5 is also called **Kanva-2.**
26. Botanical name of the Indian mulberry is ***Morus indica.***

27. RFS in **RFS-175** stands for rainfed selection.
28. Inflorescence of mulberry is popularly known as **catkin.**
29. Jassids are popularly called **leaf hoppers**.
30. The damage by jassids in mulberry is commonly called **hopper burn**
31. Incidence of mulberry leaf webber is severe during **winter month**
32. The powdery mildew disease in mulberry is caused by ***Phylloctinia corylea.***
33. **S-36 mulberry** variety y is well suited for chawki rearing.
34. Root knot in mulberry is caused by **nematode attack.**
35. **IPM** is the best for better management of mulberry pests.
36. The insect used in the biological control of tukra mealy bug on mulberry is **lady bird beetle.**
37. Mulberry seed is **oval** in shape.
38. Both adults and grubs of ***Holotrichia*** cause damage to mulberry
39. Japanese uji fly/Hime uji fly lays eggs on **mulberry leaf**
40. Mulberry has **deep root system**.
41. A major defoliating pest of mulberry is **Bihar hairy caterpillar**
42. Mulberry seeds do not have **dormancy period**.
43. **Slightly acidic to neutral soils** are ideal for mulberry growth and development.
44. Root knot disease of mulberry is caused by ***Meloidogyne incognita.***
45. Fixation of nitrogen in mulberry is by **Azotobacter**
46. Nitrogen deficiency in mulberry causes reduction in **palisade layer of leaf.**
47. Mulberry is propagated in India through: **Sapling**
48. The ancestor of *Bombyx mori* L is: ***B. mandarina***
49. In mulberry leaf the biting factor is: **Morin**
50. Second country to learn art of sericulture: **Korea.**

Model Paper: 08

1. Transovarial transmitted disease of silkworm is: **Pebrine.**
2. Green muscardine disease is caused by: ***Metarhizium anisoplea***
3. Number of ovarioles in *Bombyx mori* adult are: **08.**
4. Labin a sensitive lobe is present in: **Female silk moth**.
5. Follicular imprints are found on: **Egg**.
6. Number of prolegs in *Bombyx mori* are :**10.**
7. Number of spiracles in silkworm are: **09**.
8. Exceptional multivoltine race in Karnataka is: ***C. nichi.***
9. Japanese silkworm races spin: **Peanut Cocoon**.
10. Inner lining of foregut is called: **Intima.**
11. Spiral reinforcement structure of trachea is: **Taenidia**
12. Type of orientation of mouth parts in silkworm is: **Hypognathus**
13. Type of pupa is *Bombyx mori* is: **Obtect**
14. Cervical membrane is found in: **Larva**
15. Linulets are found on: **Second abdominal segment**.
16. Dorsal caudal horn is found on: **Dorso median line of 8th abdominal segment**
17. Peritrophic membrane is found in: **Mesenteron**.
18. Vitelline membrane is found in: **Egg**
19. Silk gland is a modified: **Labial gland**.
20. Accessory gland found is silk gland is: **Lyonnet gland.**
21. The type of setae found in silkworm *Bombyx mori* are: **Secondary hairs.**
22. Serious pest of mulberry which is commercially exploited is: ***Bombyx mori.***
23. Silk filament in the cocoon is: **Bifilar.**
24. Spun silk is manufacturing by using: **Cut and Pierced Cocoons**.
25. Number of layers of sericin is found in: **Outer layer.**
26. Amorphous form of sericin is found is: **Outer sericin.**
27. Crystaline form of sericin is found in: **Inner layer.**

28. Percent composition of fibroin in Bave is: **50%.**
29. Modification of excretory system in silkworm is: **Cryptonephridial.**
30. Red fluorescent Protein (RFC) is found in: **Haemolymph.**
31. Indians position in world raw silk production: **Second place.**
32. Unique silk present only in India in the world is: **Muga.**
33. Main function of silk exchange is: **Sale of raw silk.**
34. Biggest cocoon market in India is at: **Ramanagara.**
35. Most popular variety of mulberry in Karnataka is: **V_1.**
36. Bave comprises of: **Two brins.**
37. The process of fixing cocoon filament to the ballon is: **Casting.**
38. Mulberry is propagated through: **Seeds**.
39. The mulberry species having white stem is: ***Morus alba***
40. The mulberry species produces large fruits is: ***Morus lavigata.***
41. The type of inflorescence in mulberry is: **Catkin.**
42. Mulberry is basically a: **Tree.**
43. The plant you get from mulberry seed is: **Seedling.**
44. Mulberry plant basically has: **Taproot system.**
45. Mulberry leaf margin is: **Serrated.**
46. Victory-1 mulberry leaf is: **Entire.**
47. Tensile strength is measured in: **g/d.**
48. Elasticity of silk is represented by: **%.**
49. Size of the raw silk yarn is denoted by: **Denier.**
50. Winding test is conducted to find out: **Continuity of the yarn.**

Model Paper: 09

1. Tensile strength is measured by using: **Serigraph.**
2. Number of cottage basin are: **6.**
3. Number of ends in charaka are: **4**.

4. Which of the following is a direct reeling machine: **Charakha**
5. Which of the following reeling machine has tharpatti: **Country charakha.**
6. Chambon croissure is found in**: Country charakha**.
7. Tavelette type of croissure is found in: **Epprouvette.**
8. The circumference of the standard reel is: **1.5m.**
9. The circumference of the sizing reel: **1.125m.**
10. Number of ribs in standard reel: **6**.
11. The place of thread production in a reeling machine is called: **End.**
12. The cluster of cocoons floating in the is: **Rose.**
13. The cone shape formed by the filaments between rose and jetteboute: **Balloon.**
14. The grant reeling ratio in Improved reeling machine is: **0.656.**
15. In Japan price fixing for the cocoon is called: **Kakame.**
16. Removal of sericin from raw silk is called: **Degumming.**
17. Lengthwise yarn in a fabric is called: **Warp.**
18. Self-designing device in a loom is called: **Jacquard.**
19. Specific gravity of silkworm egg is: **1.064.**
20. Mother moth examination is done to detect: **Protozoa.**
21. The number of Rectal Papillae in silkworm rectum are: **6.**
22. Malpighian tubules arise from: **Hindgut.**
23. Specialized excretory cells are called: **Nephrocytes.**
24. Hook-like structures found in Male moth are: **Harpes.**
25. Number of genital openings in female silk moth: **2.**
26. During mating aedeagus enters into: **Bursa.**
27. Silkworm pellet is: **Hexagonal.**
28. Contraction of heart is called: **Systole.**
29. The openings of the heart are called: **Ostia.**
30. Isolation chamber is used for: **Incubation.**

31. Surface sterilization of egg is done using: **Formalin.**
32. Hydrated lime is: **$Ca\ (OH)_2$**
33. Active lime is: **CaO.**
34. Shoot feeding is best suited for: **Second and third instar**.
35. Silkworm feeds actively during: **Cool hour.**
36. Indian mulberry silkworm races are: **Tetra moulters**.
37. The silkworm *Bombyx mori* undergoes diapauses during: **Egg stage**.
38. The silkworm race NB_4D_2 is a: **Japanese race.**
39. The silkworm race CSR_2 cocoon is having: **Oval shape.**
40. In crossbreed PM x CSR2 the male parent is: **CSR_2.**
41. Acid treatment is also called as: **Hydrochlorination.**
42. Hydrometer is used to measure: **Specific gravity.**
43. The silkworm is rich in: **Vitamin B.**
44. Ecdysone activity is peak during: **Moulting.**
45. Mulberry fruit is rich in: **Vitamin C.**
46. Juvenile hormone activity is more during: **Instar.**

Model Paper: 10

1. Exogenous application of ecdysteroids hastens **Moulting Hormone.**
2. In *B. mori* Juvenile hormone is secreted by **Corpora allata.**
3. White muscardine is caused by ***Beauveria bassiana.***
4. All BmDNVs multiply in the **Nucleus of Columnar cells of midgut epithelium.**
5. Acute bacterial toxicosis is caused by ***Bacillus thurungensis.***
6. Riddling is a process of **Separation of Cocoon by size.**
7. Zero energy chamber is considered ideal for **Silkworm egg preservation.**
8. *Aspergillus awamori* is considered as **Good and Solublizer.**
9. CSR&TI is located at **Mysore.**
10. Vishala mulberry variety is recommended for: **Rainfed Conditions.**

11. Pupal diapause is very common in **Tasar Silkworm.**
12. Rectal protrusion in silkworm is due to infection of: **Bm CPV.**
13. Green muscardine is caused by ***Metarhizium anisoplea.***
14. *Bombyx mori* kenchu disease is caused by **Non- Occluded Virus.**
15. By nature, mulberry is a **Tree.**
16. Open type of cocoon is spun by **Eri silkworm.**
17. National silkworm seed project comes under **CSB.**
18. Croissure on reeling machine ensures **Uniform tension.**
19. Under 3 tier system of egg multiplication P1 seed cocoons are prepared at **Seed areas.**
20. Winnowing of eggs is commonly practiced for removal of **Unfertile eggs.**
21. The trade name of chlorine dioxide is **Sanitech.**
22. The somatic chromosomes of diploid mulberry are **2n=28.**
23. Low cost incubation chamber is developed for the purpose of **Uniform Development of Embryo.**
24. "Ushnakoti" is **Coccon drying Chamber.**
25. The book written by Dr. Y. Tazima is **Slkworm; An important laboratory tool.**
26. In silkworm fibroin is synthesized in **Posterior region of silk glands.**
27. Filipps glands are associated with **Silk glands.**
28. Silk glands open into **Spinneret.**
29. Excretory product of silkworm is **Uric acid.**
30. Single cocoon silk filament has **2 Brins.**
31. Single mother moth examination for pebrine disease is done at **P4 and P3 Stations.**
32. Which of the following cocoon has two pupae inside **Dupion.**
33. *Terminalia arjuna* is the primary host of **Tropical tasar.**
34. Single cocoon reeler is called **Epprouvette.**
35. The circumference of the standard silk reel is **1.50 m.**

36. Number of worms that can be mounted on chandrike per square foot **40-50.**
37. Silkworm haluroga' in Karnataka is referred to **Grasserie.**
38. Which of the following is a pointed cocoon **Pure mysore.**
39. *Diaphania puverulentalis* is managed with ***T. Chilonis.***
40. Headquarters of International Sericulture Commision is located in **Bangalore.**
41. *Samia Cynthia ricini* silk is also known as **Ahimsa silk.**
42. Soma and Soalu are the host plants of ***A. assamensis.***
43. Size of silk cocoons is expressed as **Number/ litre.**
44. Ecdysone is secreted by **Prothoracic gland.**
45. Blackening of spinning silkworms while on mountages is due to **Bacteria.**
46. Mulberry is **Highly Cross Pollinated.**
47. Most of the silk cocoons produced in India are unfit for reeling by **Automatic reeling.**
48. The female larva of *Bombyx mori* is having **ZW Sex chromosomes.**
49. Tasar cocoon are easily cooked by treating with **Biopril 50.**
50. Natwa is commonly employed in reeling of **Tasar Cocoons.**

Model Paper: 11

1. Serigraph test is conducted to find out **Tenacity and Elongation.**
2. Renditta" is **Quantity of Cocoons required to produce 1 kg of raw silk.**
3. The Pure Mysore breed is characterized by **Long larval duration.**
4. Application of "Sampoorna on silkworm ensures **Early and uniform mounting hormone.**
5. The length and width of Pebrine spore is **3-4.*M* in length and 1.5-2.5*M* in width.**
6. In India the silkworm germplasm bank is maintained at **CSGRC, Hosur.**
7. Size of silk cocoons is expressed as **Number/ litre.**
8. The eco race of tasar silkworm is **Daba.**

9. Biting factor present in mulberry is **Morin.**
10. Mulberry silkworm belongs to the super family **Bombycoidea.**
11. The variety of mulberry suited for rain-fed condition **S-13.**
12. Excretory system in silkworm **Crypto nephridial**.
13. Total number of legs in mulberry silkworm larva is **Thorax- 3 pairs, Abdominal 11 pairs.**
14. Total number of spriracles in *Bombyx mori* larva is **9 pairs.**
15. Type of blood circulation in silkworm is **Open type.**
16. Number of instars in tetra moulters are **5 instars.**
17. Muga silkworm is **multivoltine.**
18. Eri silkworm, *Samia cynthia ricini* is **Domesticated.**
19. West Bengal Local mulberry silkworm is **Nistari.**
20. Pure Mysore seed area is **Kunigal.**
21. In *B. mori*, Juvenile hormone helps prolong **Larval period.**
22. Popular multi-bivoltine hybrid of mulberry silkworm lately is **PM X CSR_2**
23. Chawki refers to **1 & 2 instars.**
24. GMG refers to **Government model Grainage.**
25. Peanut shaped cocoons are of **Japanese.**
26. Colored pointed cocoons are: **Multivoltine.**
27. *Borrolina* virus causes **Grasserie.**
28. Pebrine is caused by **Protozoa.**
29. Calcification is a symptom of **Muscardine fungus.**
30. Sanitech is generally: **Surface Disinfectant.**
31. For prevention of muscardine the bed disinfectant used is **Vijetha supplement.**
32. Silk filament is called as **Bave.**
33. Eri and muga culture is found in **India.**
34. Viteline membrane is found in silkworm **Egg.**

35. Peritropic membrane is found in **<u>Midgut.</u>**
36. The plant we get from vegetative propagation **<u>Sapling.</u>**
37. Lobed leaf is found in **<u>M-5.</u>**
38. Mulberry variety released exclusively for chawki **<u>S-36.</u>**
39. Bottom pruning in mulberry is done at a height of **<u>15 Cm.</u>**
40. Popular method of V-1 mulberry planting in Karnataka is **(90+150 Cm) X 60 Cm.**
41. V-1 variety of mulberry is released by **<u>CSR&TI, Mysore.</u>**
42. Shade loving mulberry variety is **<u>Sahana.</u>**
43. High moisture level in mulberry leaves is in **<u>Top leaves.</u>**
44. Carbohydrates are more in mulberry plant with **<u>Top leaves.</u>**
45. Crude fibre content is more in mulberry plant: **<u>Bottom leaves.</u>**
46. Reelable defective cocoons are reeled on **<u>Charakha.</u>**
47. Micronutrient supplementation to mulberry is by **<u>Mulberry growth promoter Seriboost, A multinutrient formulation- Poshan.</u>**
48. Dermestid beetles are pests on **<u>Silk Cocoons in Storage room.</u>**
49. Leaf spot disease in mulberry is found during: **<u>Rainy season.</u>**
50. Powdery mildew disease in mulberry is common during **<u>Winter season.</u>**

Model Paper: 12

1. Mulberry plant should have attained minimum age of **<u>6 months</u>** for using it as planting.
2. Single Pure Mysore' DFL contains an average **<u>400-500 eggs.</u>**
3. Bivoltine under natural condition without acid treatment can be reared **<u>Twice in a year.</u>**
4. Voltinism refers to **<u>No. of Generations.</u>**
5. One important pupal parasitoid that can be utilized for biological control of Uzi fly is ***<u>Nesolynx thymus.</u>***
6. The shoots selected for preparation of mulberry cuttings for planting must be **<u>6-8 months old.</u>**

7. Mulberry plants are bottom pruned **Once in a year.**
8. The most ideal system of mulberry planting for rainfed condition is **Pit system.**
9. *Phyllactinia corylea* causes **Powdery mildew.**
10. The colour of silk cocoon is determined by the **Pigments Present in Sericin.**
11. When the silk fibre is impregnated with water, strength of the silk is **Decrease by 20%.**
12. The characteristic symptom of BmCPV disease of silkworms is b Swollen intersegment is **Rectal protrusion.**
13. Dithane M-45 can be used as: **Rearing bed Disinfectant.**
14. The most commonly used mountages by farmers in Karnataka are **Bamboo mountages.**
15. Group of mulberry varieties recommended exclusively for irrigated condition are **S-36, V-1, S-54, G-2.**
16. Mulberry leaf rust is caused by ***Peridiospora mori.***
17. Silkworm rearing method that requires less space is: **Shelf rearing.**
18. Red muscardine in silkworm is caused by ***Sporosporella uvella.***
19. All BmDNVs multiply in the **Nuclei of Columnar cells.**
20. Vishala mulberry variety is recommended for **Rainfed.**
21. Croissure process of reeling is useful for **Uniform tension.**
22. Under 3-tier system of egg multiplication, P1 seed cocoons are prepared at **Seed areas.**
23. A fungal disease, which produce kojic acid upon infection is **Brown muscardine.**
24. In silkworm, endomitosis occurs in **Silk glands.**
25. Evenness test for silk fibre is done on **Seriplane.**
26. "Thatte Roga" of silkworm is a **Mixed infection.**
27. Kakame is **Cost of Cocoons.**
28. Compared to late age silkworm larvae young age larvae(chawki) require **High temperature and RH.**

29. Red fluorescent protein having anti-viral activity is found in the **Digestive juices of Silkworms.**
30. Paralysis followed by death in silkworm is a common symptom of ***Bacterial toxicosis.***
31. The nucleic acid type of Bm NPV is **Double stranded DNA.**
32. The midgut becomes chalky white when silkworm is infected by **Bm CPV.**
33. Leaf roller incidence is very serious on mulberry during **June to November.**
34. Generally tropical silkworm races are **Tetra moulters.**
35. The best method of propagating mulberry is through **Stem cuttings.**
36. Seri boost is used in sericulture to increase Silkworm growth **Leaf Yield.**
37. In mulberry P deficiency causes **Inter vienal Chlorosis of Older leaves.**
38. Geographical classification of mulberry silloworm breeds is based on the enzyme activity of **Amylase.**
39. Spraying of 1% benzoic acid effectively kills Uzi fly **Eggs.**
40. Strength of the formalin to be mixed with charred paddy husk for treating fifth instar silkworms against **0.8%.**
41. The quickest method of developing an improved variety: **Introduction.**
42. Popular mulberry variety V_1 is obtained from cross between **S-30 X Ber 776.**
43. $BC_2$59 is the mulberry variety developed by **Back cross method.**
44. Mass selection in mulberry is done based on **Phenotype.**
45. Mejankari silk is from **Muga Silkworm.**
46. Biting factor present in mulberry: **Morin.**
47. Haemolymph of silkworm becomes turbid due to infection by **Bm Cpv.**
48. Lunnules in *Bombyx mori* larva are found on **2nd Abdominal segments.**
49. Chawki worms require tender mulberry leaves as they are rich in **Carbohydrates.**
50. Mulberry foliar spray developed by CSR&TI very recently **Poshan.**

Model Paper: 13

1. Silk powder is used as an ingredient in production of **Cosmetics.**
2. For detecting pebrine spores, sample size of mother moths in grainages is **20%**.
3. Silkworm breeds of European origin spins Peanut shaped cocoons **Long oval**.
4. In Karnataka, mulberry is mainly grown in **Red loamy**.
5. **Triacontanol** use in mulberry increases Leaf yield and Quality.
6. Considering the world silk scenario, India is **Second largest** Producer and the **Largest** Consumer of silk.
7. Size of silk filament is indicated by **Denier**.
8. India exports Mainly **finished silk products**.
9. Contamination in silkworm rearing bed can be minimized in **Shelf rearing**.
10. Maximum uniformity in cocoon spinning can be obtained in **Rotary mountages**.
11. Which of these chemicals is not a silkworm bed disinfectant: **Sanitech**.
12. The name "Pebrine" was coined by **De quadrefageus**.
13. Aspergillosis is most prevalent in **Japan**.
14. **Herold's gland** is present in silkworm Male larvae.
15. **Peduncle type** of cocoon is spun by Muga and tasar.
16. Raw silk is also called as **Grege**.
17. Mulberry absorbs all major nutrients when soil pit is around **6.5-7.5**.
18. During Oogenesis, the stage of Meiosis at which **"Synaptonemal Complex"** is formed Pachytene.
19. Austrov's method of parthenogenesis induction, the time required for heating unfertilized eggs at 46°C is **18 mins**.
20. Silk was first discovered in: **China**.
21. Mulberry grows well with a temperature range of: **22 - 30°C**.
22. For mulberry cultivation the acidic soil can be rectified by the application of: **Lime**.

23. The annual mulberry leaf yield (in kg) from one hectare of land under rainfed condition is: **12,000 - 15,000**.
24. Powdery mildew in mulberry is caused by: **Phyllactina corylea**.
25. In mulberry silkworm, diapause occurs in: **Egg stage**.
26. The hibernating eggs are preserved in: **Cold storage**.
27. In silkworm rearing houses the instrument – hygrometer is used for measuring: **Humidity**.
28. At the time of emergence, moth excrete: **Alkaline fluid**.
29. The optimum humidity for silkworm moulting: **65 - 70 %.**
30. Uzi fly belongs to the family: **Tachinidae**.
31. The important by product of grainages is: **Pierced cocoons**.
32. Pierced cocoons are used to get: **Spun silk**.
33. The following mulberry variety is the highest yielder: **Victory - 1**.
34. The raw silk imported from China is: **2A - 3A**.
35. Mulberry fruit is: **Sorosis**.
36. Mulberry is: **Highly cross pollinated**.
37. Tukra symptoms in mulberry caused by: **Maconellicoccus hirsutus**.
38. Hand collection of egg masses and young larvae is suggested for the control of: **Diacrisia oblique**.
39. Safety period suggested after spraying DDVP is: **15 - 17 days**.
40. Thrips infestation in mulberry is identified by: **Brown presence of white streakes which become yellowish**.
41. White roots rot of mulberry is caused by: **Rosellina necatrix**.
42. Acedrum mori is responsible for the infection of: **Leaf rust**.
43. The total moulting time required at different instars of the silkworm is: **15 to 30 hours**.
44. The excretory product of silkworm is: **Uric acid**.
45. Black scars are found on silkworms infested with : **Uzi fly**.
46. Temperate climate is suitable for the rearing of these silkworms : **Univoltine**.

47. The purpose of cocoon stifling is : **To kill pupae**.
48. The average number of eggs in a DFL is : **450 to 500**.
49. Application of CAN in mulberry is found beneficial in : **Acidic soil**.
50. Fertigation efficiency is more in : **Drip irrigation**.

Model Paper: 14

1. Application of green manure and vermicompost to mulberry garden increases : **Organic matter content of the soil**.
2. This nutrient becomes quickly available to plants : **Ammonical form of nitrogen**.
3. The variety is found best for chawki rearing: **S_{36}**.
4. All commercially exploited mulberry varieties are: ***Morus indica***.
5. The silk worm respires through: **Tracheae**.
6. The cross involving the F_1 hybrid and the potential double recessive is called : **Tri - hybrid cross**.
7. Equatorial plate is formed during: **Metaphase**.
8. Diapause in *Bombyx mori* is caused by: **Hormone**.
9. The nature of chromosome in silkworm is: **Holocentric**.
10. 2n number of chromosomes in silk worm is: **28**.
11. The chromosome which play a decessive role in determination of sex in *B. mori* is: **W chromosome**.
12. The silk glands are derived from: **Ectoderm**.
13. The major nitrogenous excretory product in silkworm is: **Uric acid**.
14. Renditta is: **The quantity of cocoons in kg required for 1 kg of silk**.
15. Which of the following is required for hybridization in mulberry plants: **Emasculation**.
16. Amylase hydrolyses: **Starch**.
17. The inheritance of cocoon weight is controlled: **Polygenically**.
18. Auto sexing in silkworm is achieved through: **Translocation**.
19. The method of breeding that has contributed for increasing silk productivity is: **Heterosis breeding**.

20. The industrial silkworm is usually: **Hybrid**.
21. Cocoon cooking and reeling is performed in same basin in : **Country charaka**.
22. The number of reeling ends per basin in multi end reeling machine is: **8 – 10**.
23. In India temperate sericulture is in: **Jammu and Kashmir**.
24. Silkworm Seed Technology Laboratory is located at: **Bangalore**.
25. Transovarially transmitted disease in silkworm is: **Pebrine**.
26. The non - soluable silk protein is: **Fibroin**.
27. *Aspergillus* is a: **Fungal disease**.
28. Female *Bombyx* pupa is identified by its 'X' mark on the ventral side of segment: **8th**.
29. The scientific who discovered pebrine: **Louis Pasture**.
30. Bivoltine eggs are to be cold stored within how many hours of egg laying: **48hours**.
31. Excess of water from silk bale during reeling is removed by: **Croissure**.
32. Calcified cocoons are the one which have been destroyed by: **Fungus**.
33. The denier of pure Mysore: **2**.
34. Dupion silk is obtained from : **Double cocoons**.
35. The highest quantity of raw silk produced in: **China**.
36. Basic seed means: **Parental seed**.
37. The standard number of cocoons per kilogram in pure Mysore: **900 – 1000**.
38. The hierarchy of basic seed organization is in the order: **P_3, P_2, P_1**.
39. Dusting of sulphur is a measure suggested for the control of: **Powdery mildew**.
40. Appearance of small water soaked irregular spots on the lower surface of leaves in a characteristic symptom of: **Bacterial leaf blight**.
41. Appearance of root knots in mulberry is due to: **Nematodes**.
42. Marginal scorching and tip burning are a characteristic deficiency symptom of: **Potassium**.

43. Stem turning woody and short is due to the deficiency of: **Magnesium**.
44. The average number of eggs laid by the multivoltine female moth is: **400**.
45. In total cost of silk production, the per cent cost of mulberry leaf is: **60**.
46. Idioblast in mulberry leaf content: **Calcium carbonate**.
47. Highest chromosomal number is seen in: ***Morus nigra***.
48. The larval reproduction markings in female silkworm is: **Ishiwata's gland**.
49. Jaundice disease in silkworm is also called as: **Grasserie**.
50. The surface layer of cocoon is called: **Floss**.

Model Paper: 15

1. Leaf preservation in necessary for: **To maintain freshness of the leaf**.
2. The whole shoot harvest is very common in: **Irrigated mulberry**.
3. The state in India, which is most suitable for rearing univoltine silkworm is: **Jammu and Kashmir**.
4. Extension aims to change: **Knowledge, skill and attitude**.
5. Silkworm males are: **Homogametic**.
6. Ecdysone is responsible for: **Moulting**.
7. The mouth parts of silkworm are of: **Mandibulate type**.
8. Adhesive or gluey substance is produced from: **Accessory glands**.
9. The mulberry silk can be reeled through: **1. Charaka 2. Cottage basin 3. Filature.**
10. Glands that discharge their products directly into blood stream are called: **Endocrine**.
11. The process of removal of metabolic wastes from the body is called: **Excretion**.
12. Two forms of the same gene are called: **Alleles**.
13. Denier denotes: **Thickness of the filament**.
14. The all-round development of silk industry in India is in the charge of: **CSB**.

15. The effective method of irrigation involving less wastage of water in: **<u>Drip irrigation</u>**.
16. The Crescents in silkworm are located on the abdominal segment: **<u>Second</u>**.
17. The basic structure of nervous system is: **<u>Neuron</u>**.
18. The appropriate chemical disinfectant used to disinfect rearing appliance in open condition: **<u>Bleaching powder</u>**.
19. Frequency of feeding silkworm depends on: **<u>1. Method of feeding 2. Season of rearing 3. Method of rearing</u>**.
20. Double cocoon formation is largely due to: **<u>High density of mounting</u>**.
21. Juvenile hormone is secreted by: **<u>Corpora allata</u>**.
22. The disease caused by non-occluded virus is: **<u>Densonucleosis</u>**.
23. The variety of mulberry evolved through mutation using EMS is: **$\underline{S_{36}}$**.
24. E-group of alleles are located on: **<u>Sixth chromosome</u>**.
25. White muscardine is severe during: **<u>Winter</u>**.
26. Loose eggs are preferred because of : **<u>Quantification and high fertility</u>**.
27. Tensile strength of raw silk is tested using: **<u>Seriplane</u>**.
28. Chambon croissure system is followed in: **<u>Charaka</u>**.
29. Evenness of raw silk is the prime objective of reeling in: **<u>Automatic reeling machine</u>**.
30. The optimum space between the spiral in a skein is: **<u>5 cm</u>**.
31. Chromosomal formula for female silkworm is: **<u>ZW</u>**.
32. Renditta for cross breed silkworm is: **<u>Eight</u>**.
33. To produce one kilogram of raw silk the requirement of mulberry leaves is: **<u>20 kg</u>**.
34. The key amino acid for silk synthesis is: **<u>Glycine</u>**.
35. Biting action in silkworm towards mulberry leaves is triggered due to the presence of: **<u>Morin.</u>**
36. Early detection of silkworm pathogens can be done by: **<u>ELISA</u>**.
37. White muscardine appears at a humidity of: **<u>80 – 90 %</u>**.

38. Per hectare employment generation in sericulture is: **4,000-man days**.
39. Largest organ in silkworm is: **Digestive system**.
40. The defective cocoons can be reeled by: **Charaka**.
41. Per day raw silk production in cottage basin is: **600g**.
42. Hopper burn is a typical symptom of: ***Empoasca flavescens***.
43. The powdery mildew disease of mulberry is more revelent in: **Closely planted gardens**.
44. The dumbbell shaped cocoon in a characteristic trait of: **Japanese races**.
45. One kilo gram of seed cocoons of *Bombyx mori* (about 600 members) can yield approximately: **100 DFLs**.
46. *Beauveria bassiana* causes: **White muscardine**.
47. The latest valid scientific name of mulberry silkworm uzifly is: ***Exorista sorbillans***.
48. In shoot feeding method of silkworm rearing, leaf can be served up to: **20 – 30 %**.
49. Mulberry variety with succulent, glossy and leathery leaves is: **Victory - 1**.
50. Skeletinization of mulberry leaf is a characteristic symptom of: **Early age caterpillars of *Spodoptera litura***.

Model Paper: 16

1. Sanitech is an ideal general disinfectant containing: **Chlorine dioxide**.
2. Through the incidence of uzi fly on silkworms is noticed throughout the year, but its incidence is maximum during: **Winter months**.
3. The oval shaped cocoon is a characteristic feature of: **Chinese breeds**.
4. Cocoon riddling is process in which the cocoons are separated according to: **Cocoon size**.
5. The protein content in the raw / fresh pupae of *Bombyx mori* is: **40 %**.
6. The V_1 variety of mulberry is developed by: **CSRTI, Mysore**.
7. The quality of the silk yarn of CSR hybrids is: **2A - 3A grade**.
8. The Kenchu disease of silkworm is generally common during: **Summer months**.

9. The yellow muscardine disease of silkworm is caused by: ***Isaria farinose***.
10. The predisposing factor for brown muscardine disease of silkworm are: **Low temperature and high humidity**.
11. The recommended quantity of FYM/ compost for irrigated mulberry garden per hectare per year is: **20 MT**.
12. The scientific name of leaf roller pest of mulberry: ***Diaphania pulverulentalis***.
13. The sex- limited CSR hybrid is: **CSR_{18} x CSR_{19}**.
14. The optimum light required for healthy silkworm growth and development is: **20 – 30 lux**.
15. The circumference of the standard re - reel is: **1.50 m**.
16. The standard *Bombyx mori* L. loose egg pack contains: **20,000 eggs**.
17. Visual inspection of raw silk involves which of the following tests: **1. Evenness 2. Neatness 3. Cleanness**.
18. In mechanical spinning machine basically have: **Wheel and pedal**.
19. ARM stands for: **Automatic Reeling Machine**.
20. Internal suture is made up of: **Silk gland**.
21. Rearing waste can be used for production of: **1. Compost and vermicompost 2. Biogas 3. Mushroom production**.
22. For management of Papaya Mealy Bug predator used is : ***Acerophagous papaya***.
23. Trenching and Mulching in mulberry garden improves: **1. Soil fertility 2. Moisture conservation 3. Weed management**.
24. Recently CSRTI Mysore developed mulberry foliar spray: **Poshan**.
25. Bio agent used for the management of leaf roller in mulberry: **T. chilonis**.
26. OFIT: **One feed per instar**.
27. CPP: **Cluster Promotion Programme**.
28. SCTH Stands for: **Silk Conditioning and Testing House**.
29. Bio-Craftts helps in better use of: **Defective Cocoons**.
30. Vishala mulberry variety is recommended mainly for: **Rainfed**.

31. ISC Head quarter is in **India**.
32. Concept of Chawki rearing centre helps farmers in: **Improving cocoon field , Saves time and Maintains hygiene**.
33. Sericin used in production of cosmetics like: **Face cream**.
34. Pupal powder is used in production of: **Biscuits**.
35. Fixing of nitrogen in mulberry: **Azatobacter**.
36. *Maconellicocus hirsutus* is: **Mealy bug**.
37. Mulberry variety resistant to powdery mildew disease is: **MR - 2**.
38. Main symptom of *Spilosoma obliqua* infection is: **Leaf skeletonization**.
39. Mulberry species raised as tree in 17K for production of fruits is: ***Morus multicalis***.
40. Major item of Indian silk export is: **Finished fabric**.
41. Silk filaments is technically known as: **Bave**.
42. The size of the cocoon is expressed as: **Number/litre**.
43. Flooding is one of the control measures for: **Leaf roller**.
44. Mulberry inflorescence is called: **Catkin**.
45. Most suitable soil type for mulberry is: **Red loam**.
46. The key amino acid for silk synthesis is: **Glycine**.
47. Uzi fly is a: **Regular pest**.
48. Uzi maggots have: **3 instars**.
49. Brin consist of: **1 fibroin + 3 sericin layers**.
50. The raw silk is packed as: **Bale**.
51. The circumference of the charaka reel is: **150 cm**.
52. The standard PH of the boiling water during reeling is: **8.6**.
53. Madhya Pradesh is first in: **Tasar silk production**.
54. Chawki worms can be indented after: **50-55 days of pruning in summer**.
55. Cellular rearing is done in: **P_4 and P_3 station**.
56. Eggs attain pin head stage: **48 hours before hatching**.

57. Chawki worms require the leaves with more: **Carbohydrates**.
58. Number of worms mounted on Chandrika per square feet: **40 - 50**.
59. Life span of muscardine spores: **5 years**.
60. Number of worms required to rear 100 dfls multivoltine chawki worms: **50**.
61. The number of spermatozoa in a male *Bombyx mori* L. is around: **1.54×10^6**.
62. Pure silk in fabrics can be identified by **Silk Mark** with silk moth motif on **Golden Yellow** tag given by **SMOT** silk mark organization of India.